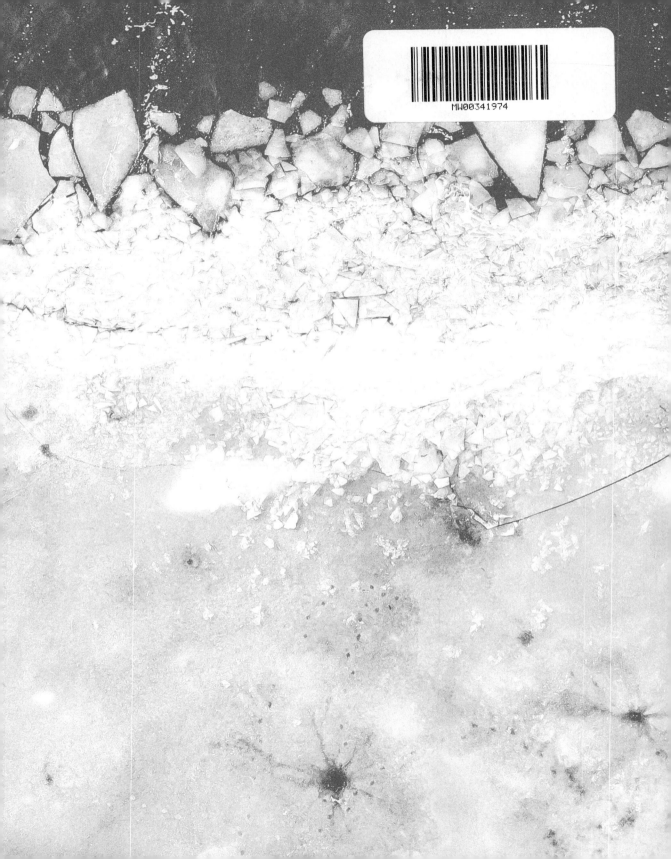

The Human Age

To my grandchildren,
Gísli Þór, Jón Bjarni, Saga Rós
and Úlfur Bergmann.

Published in 2020 by Welbeck, an imprint of the Welbeck
Publishing Group), 20 Mortimer Street London W1T 3JW

Text © Gísli Pálsson
Design © Welbeck

A CIP catalogue record for this book is available from the
British Library

ISBN 978 1 78739 435 3

Printed in Dubai

10 9 8 7 6 5 4 3 2 1

The Human Age

How we created the Anthropocene
epoch and caused the climate crisis

∙∙

GÍSLI PÁLSSON

WELBECK

CONTENTS

· · · · · · · · · · · · · · · · · · · ·

"It is man's Earth now. One wonders what obligations may accompany this infinite possession."

Henry Fairfield Osborn Jr, *Our Plundered Planet*, 1948

....

"What used to be called nature
has erupted into human affairs,
and vice versa, in such a way and
with such permanence as to change
fundamentally means and prospects
for going on, including going on at all."

Donna Jeanne Haraway, *Staying with the Trouble*, 2016

....

"The *Anthropocene* defines
Earth's most recent geologic time
period as being human-influenced,
or anthropogenic, based on
overwhelming global evidence ..."

Welcome to the Anthropocene web portal, www.anthropocene.info, 2019

The Economist

Obama, Bibi and peace

Britain's privacy mess

The costly war on cancer

How the brain drain reduces poverty

A soft landing for China

MAY 28TH–JUNE 3RD 2011

Economist.com

Welcome to the Anthropocene

Geology's new age

Introduction: A new epoch

In the summer of 2011, the cover of *The Economist* presented a striking image of Earth roaming in dark space, with the heading "Welcome to the Anthropocene: Geology's new age". The image showed Earth covered with metallic plates, bolts and nuts. Some plates were badly damaged, and through the cracks it was possible to see parts of the interior, with a hint of a glowing furnace. This planet was clearly crafted by humans, from top to bottom, and it appeared to be warming up fast. This image spoke to a great truth: in recent decades, life on planet Earth has been refashioned on an unprecedented scale through the expansion of human activities. Understandably, many people find it necessary to put a name to this new age – the Anthropocene. Some have envisioned a "good Anthropocene", a space for hope; perhaps global warming would provide new opportunities, including warmer conditions in traditionally cold regions and the opening of the Northwest Passage through the Canadian Arctic Archipelago, between the North Atlantic and the Pacific Ocean – a recurring European dream for centuries. For others, the new epoch would be necessarily catastrophic, ugly and dangerous.[1]

Perhaps there was a hint, too, of arrogance in *The Economist*'s provocatively jaunty headline, a

Above: Polar bears use Wrangel Island, Russia, as a summer resting location while the sea ice has melted. Climate breakdown means the bears are now spending an unprecedented amount of time on land rather than on ice, away from their primary food sources.

Opposite: Cover image of *The Economist*, 28 May–3 June, 2011.

threatened divide between inviters and invitees? Who were the imagined guests? What kind of festivity, if that was the right word, might be involved? Given what we now know about the scale of the environmental hazards of the Anthropocene, the invitation seems ill-judged, even distasteful; for comparison, one might consider a headline reading "Welcome to the Holocaust".[2] The figurative invitation to the Anthropocene can be read in several ways. To most readers of *The Economist* in 2011, the term "Anthropocene" was new, demanding explanation. By now, it seems ubiquitous, in the media and across many fields of scholarship, literature and the visual arts.[3] Googling the term in February 2020 produced more than seven million hits, a number that continues to increase every day.

The word "Anthropocene" combines the root "anthropo-", meaning "human", with the root "-cene", the standard suffix for "epoch" in geologic time. The human impact implied by the term, like the traces of earlier epochs in geologic history, is recorded in the earth itself and in our bodies, and manifests in a variety of ways. The globe displayed at a climate protest in Italy in 2019 (pictured) was very different to that of the magazine cover; the Anthropocene was now unmistakably bad, the world up in flames. The image and reputation of the Anthropocene had changed radically in only eight years. Global warming had become global heating, a road to total disaster paved by humans.

There is much more to the story of the Anthropocene and it is vital for our future, the fate of humanity and of life itself, that we are well informed, seriously concerned and ready for action. This book briefly explains how and why the term arose, what it means, how it generates debates and denials, why it is flawed and why

many people nevertheless find it useful, even central, to facilitate human engagement with current environmental hazards. While the notion of the Anthropocene was initially conceived simply as a technical term in geoscience, a means of labelling or categorizing a new geological age, it has refused to accept confinement. It has become a concept that collapses natural and social history into one another. Not only have we become a dictating factor, we have become part of *geos*, the Earth itself. Everything is both geologic and human at the same time.

Moving beyond the concept itself to the damaged Earth, the book explores particular Anthropocenic changes – notably weather extremes, the melting of glaciers, the dying ocean and plastic that refuses to disappear – and their implications and challenges for everyday life in specific contexts. This is not an encyclopaedia on everything Anthropocenic – that would take dozens of volumes; rather, the point is to present the reader with key cases and themes, followed by visual material, in order to illuminate our new place in the modern world and the precarious state of the planet. Among the most pressing concerns is the sixth great wave of extinction in the history of the planet. Humans are not the only personae in this saga, and accordingly the book addresses a series of other key players – including animals, plants, microorganisms, mountains, oceans and wetlands. The Anthropocene, by definition, necessitates broadening our gaze far beyond humanity to the Earth "itself".

The extinction of species and the resultant loss of biodiversity, one of the signatures of the Anthropocene, has important implications for what is now called "planetary health". The Covid-19 pandemic in the wake of the global spread of the

Above: The "Bad Anthropocene": Climate protest in Italy, 2019.

coronavirus in late 2019 and early 2020, with its devastating impact on public health, international travel and the economy, was not a purely "natural" catastrophe but the by-product of human activities. Such epidemics, it has become increasingly clear, are driven by human refashioning of the planet, its habitat and its life forms, "shaking the viruses loose"[4] from their hosts. At the time of writing, the Anthropocenic relevance of the Covid-19 epidemic remains relatively subdued in public media, as governments and the general public busily grapple with the spread of the virus and its other potential outbreaks of the kind in the near future. Within weeks, however, the slowdown and halting of traffic in the wake of the virus became a healthy reminder of the scale of the carbon footprint of human travel. Air pollution in some of the most congested cities, such as Beijing, was quickly reduced. At the same time, the canals of Venice became cleaner as tourism collapsed.[5]

Given the dramatic developments captured by the idea and the realities of the Anthropocene, it is vital to reflect on the history of human environmental relations, prospects for the future and the opportunities for action. This book explores human attempts to come to terms with the Anthropocene, and to mitigate or reverse its damaging environmental and social impact. This involves, crucially, focusing on the spaces for hope and action, especially for meaningful solidarity at all levels. If this spectacular, urgent and timely endeavour proves to be unsuccessful, the dawn of the young Anthropocene will also mark its end – the termination of human history.

Gísli Pálsson, Reykjavík, May 2020

Prelude

Challenges to the Anthropocene

Just when the Western world had concluded that technology and science could solve every problem, storm clouds began to gather. From around the middle of the last century, a lively debate began about the planet's limitations. In 1979, Swiss writer Max Frisch observed in *Man in the Holocene* that novels are no use for "days like these, they deal with people and their relationships ... with society, etc., as if the place for these things were assured, the earth for all time earth, the sea level fixed for all time".[1] A perceptive artist, Frisch seemed to anticipate what was coming, although he had no name for "days like these". No one spoke of a new era until 1988, when environmental historian Donald Worster published *The Ends of the Earth*. Worster suggested humanity was "approaching a grand ritualistic climax". He wrote: "It is irresistible to ask whether we are passing from one era into another, from what we have called 'modern history' into something different and altogether unpredictable".[2] That "something different" is the Anthropocene.

Concerns with the state of the Earth are not new, though. As early as 1864, George Perkins

Previous spread: Ice loss near Ilulissat, Greenland, summer 2019.

Above: Paria Canyon-Vermilion Cliffs Wilderness, Arizona, USA, showing Jurassic sandstone rock formations.

Opposite: Geologic timescale, 650 million years ago to the present.

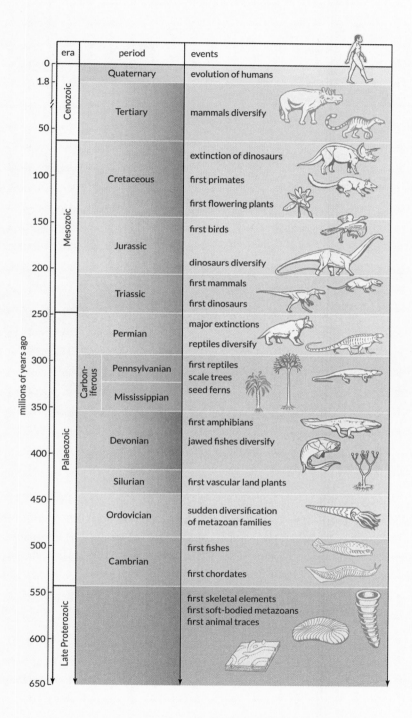

era	period	events
Cenozoic	Quaternary	evolution of humans
	Tertiary	mammals diversify
Mesozoic	Cretaceous	extinction of dinosaurs first primates first flowering plants
	Jurassic	first birds dinosaurs diversify
	Triassic	first mammals first dinosaurs
Palaeozoic	Permian	major extinctions reptiles diversify
	Carboniferous — Pennsylvanian	first reptiles scale trees seed ferns
	Carboniferous — Mississippian	
	Devonian	first amphibians jawed fishes diversify
	Silurian	first vascular land plants
	Ordovician	sudden diversification of metazoan families
	Cambrian	first fishes first chordates
Late Proterozoic		first skeletal elements first soft-bodied metazoans first animal traces

millions of years ago

0
1.8
50
100
150
200
250
300
350
400
450
500
550
600
650

Marsh suggested that it was "fast becoming an unfit home for its noblest inhabitants".[3] It was only quite recently, however, that a substantial number of people began to see that something was seriously wrong with our planet. Now, with current discussions of the Anthropocene, there is a general sense that humans are faced with a critical and narrowing window of opportunity; if we do not act very soon, it may be too late. The concept of the Anthropocene is one of the most recent and influential of several concepts attempting to capture the reality that humans have become a dominant factor in shaping nature.[4]

Concepts similar to the Anthropocene date back as far as 1873, with Antonio Stoppani's "Anthropozoic era", while more recent iterations include Andrew Revkin's "Anthrocene" (1992) and Michael Samways's "Homogenocene" (1999). The earliest hint of the idea that humans might be altering the foundations of the Earth was probably the seventh and final epoch in the Comte de Buffon's scheme, "When the power of man has assisted that of nature", outlined in his book *The Epochs of Nature*, originally published in 1778 (although it did not appear in English until 2018).[5] While the notion of the Anthropocene has been fairly broadly accepted, it has also been the subject of constant, heated debate. One of the debates concerns the origin of this new era, and when and how it began to take effect.

In one of the more popular versions of the Anthropocene, it is conceived of as having begun around the middle of the last century, with the advent of nuclear weapons and energy, which would threaten life as we know it and leave permanent traces on the planet in the form of radioactive material. For others, a "great acceleration" in human activities, and their impact, began with the Industrial Revolution and the massive extraction and exploitation of fossil fuels from the second half of the eighteenth century, resulting in an exponential increase in the levels of CO_2 and other greenhouse gases in the atmosphere. After all, these were the developments that produced many of the Anthropocene's most troubling symptoms.

Another way of looking at the Anthropocene derives from the insight that it is the first geological epoch in which the defining geological force is conscious of its geological role.[6] From this perspective, the Anthropocene commences in its fullest sense at the point when human awareness of our role in shaping the Earth began to affect our active relationship with the environment. Under this reading, it becomes perhaps contentious to speak of a geologic epoch. It may make more sense to refer to a new era in human–environmental relationships. The use of the term "Anthropocene" itself can be understood as a signal of this new relationship.

The idea of the Anthropocene was coined by Dutch chemist Paul Crutzen at a scientific conference in 2000. Irritated by the preoccupation at the conference with the Holocene (the "entirely recent" epoch that began at the end of the last ice age 11,500 years ago), Crutzen blurted out: "Let's stop it! We are no longer in the Holocene. We are in the Anthropocene". This was a flash of insight, and within minutes, the term "Anthropocene" became the main topic of conversation.[7] Crutzen's point, of course, was that the comparatively recent onset of massive changes to the planet, with the potential for huge implications for all life on Earth, deserved specific recognition in the scheme of geologic time.

Since Crutzen's dramatic intervention, there has been consideration of giving formal status in geological classification systems to the term "Anthropocene". In 2008, the Stratigraphy Commission of the Geological Society of London indicated that there was merit in considering the possible formalization of the term: adding it eventually to the Cambrian, Jurassic, Pleistocene and other such units on the geological time scale.[8] Consideration of the Anthropocene as an epoch, it was argued, seemed reasonable. The scale of change that had taken and was continuing to take place appeared to have taken the Earth out of the boundaries of conditions and properties that mark the Holocene epoch; the Anthropocene, as a result, represents a new phase in the history of both humankind and of the Earth.

The idea of allocating the Anthropocene a respectable place in the global standard of geologic periods has, however, been contested. Geologists have actively debated the stratigraphic "legalities" of the Anthropocene, arguing about whether the term would meet their strict professional protocols, how to detect the relevant signatures or "spikes" in the geological strata (layers of rock),

Opposite: Germany closes its last coal mine, Tagebau Garzweiler near Cologne, 2019.

CHALLENGES TO THE ANTHROPOCENE

and so on.[9] These spikes, critics have argued, are barely visible in the geologic record.

In July 2018, the International Commission on Stratigraphy announced, after long and winding proceedings, that the Holocene epoch should be split into three subdivisions: the Greenlandian, the Northgrippan and the current Meghalayan Age. The last one of these, the Commission specified, began with a 200-year drought that affected civilizations around 4,000 years Before Present, leaving open the possibility of an Anthropocene era starting from 1950. We need not dive into the details of the peculiar names and identities of the three Ages. For many people, including some geoscientists, the Commission's statement indicated that the geosciences were out of touch with both geologic and social strata.[10] While physics, and even popular culture, has deconstructed and redefined time and space as relative categories, geoscientists tend to treat geologic periods as if they are carved in stone, much like the Early Modern classification of living species, in the fashion of Carl von Linné's grand, inert "System of Nature".

It is true that the Anthropocene represents a departure in the discourse of the geosciences, in the sense that, unlike the concept of the Holocene, it is future-orientated, focusing on spikes that still haven't firmly placed themselves in geologic strata (leaving aside, perhaps, plastic, chicken bones, and radioactivity). Yet our historical signatures can be undeniable and telling, especially mass extinction. Periodic divisions and the conventions for naming ages and chronotypes depend, much like personal names, on a community of speakers for their success and survival. Some labels are momentary nicknames, so to speak, quickly relegated to history, while others "stick", granted the licence of a community. It is clear that the concept of the Anthropocene, and the accompanying label, have gained traction in the public domain, despite the continued scepticism and occasional hostility of many geoscientists and social theorists.[11] For a rapidly growing number of people, it serves as a powerful instrument of discourse, directing public attention to the biggest challenges ever faced by humanity, and the responsibilities these entail.

LES ÉPOQUES
DE
LA NATURE,
PAR MONSIEUR
LE COMTE DE BUFFON,
Intendant du Jardin & du Cabinet du Roi, de l'Académie Françoise, de celle des Sciences, &c.

TOME PREMIER.

A PARIS,
DE L'IMPRIMERIE ROYALE.

M. DCC. LXXX.

Opposite: The footprint of the Anthropocene.

Above: The cover of the Comte de Buffon's book *The Epochs of Nature* (1778).

THREE

The recognition of deep time

For a long time, Western scholars generally believed that the Earth was a flat disc devoid of any history stretching back more than a few thousand years. Some ancient Hellenic philosophers (including Eratosthenes), and no doubt some non-Western theorists and cosmologies through the ages, did develop the idea of a globe – even making calculations of circumference – but for one reason or another, these ideas generally did not survive in Western discourse. The earliest surviving representation of the Earth as a globe dates from 1492. The conception of the Earth as a sphere naturally led to questions about its internal nature and its history. Sometimes its interior was described in terms of arteries and entrails, all in frenzied motion. The later development of the concept of what is now known as "deep time", the name of which was coined by American writer John McPhee in 1981, marked a profound turning point in the understanding of life, the material world, and their long-evolving shared history.[1] At the same time, it facilitated human exploitation of the fabric of the Earth and

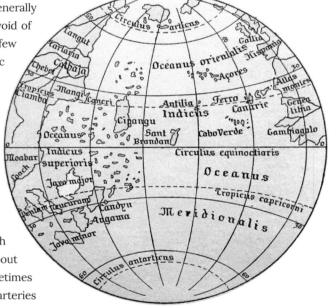

the extraction of resources on an unparalleled scale, leading to the current Anthropocene.

Eventually, the subterranean world was carefully mapped, providing a glimpse of the deep layers below the surface of Earth and the history they might disclose. The first geologic map, now preserved in a museum at the University of

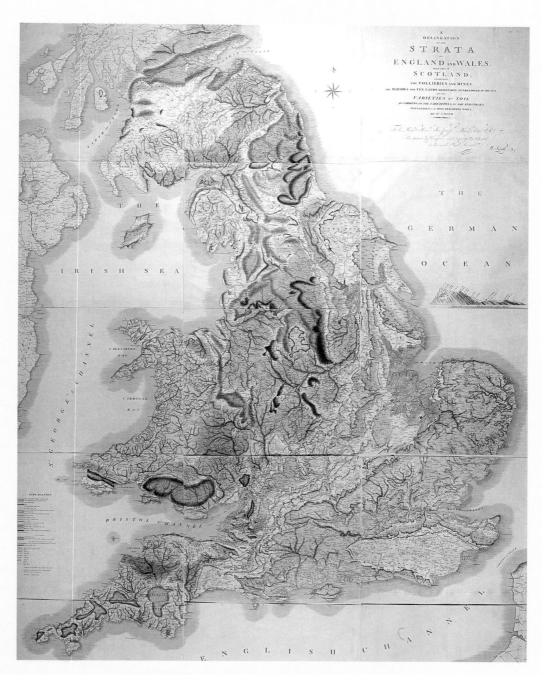

Opposite: The earliest known image of the globe, Martin Behaim, 1492.

Above: The first geologic map. William Smith, 1815,

THE RECOGNITION OF DEEP TIME

MARY ANNING

Anning grew up in a poor household. Her father, Richard, was a carpenter, and the family's income was mainly derived from selling fossils to travellers and naturalists. Fossil collecting was both a popular pastime and a scientific endeavour, and many scientists would correspond with and visit Mary Anning. The mysterious animals discovered by Anning had a huge impact on scientific understanding of the history of the planet and the animal kingdom. The French-American biologist Louis Agassiz often sought Anning's advice, even naming some "new" fossils after her, such as *Acrodus anningiae* (an extinct cartilaginous fish). But few visitors would mention Anning in their writings. "The world has treated me badly," she said, "These learned men have sucked my brain." Anning's indignant comment draws attention to the importance of observing planetary observers themselves, all of whom are necessarily situated in time and history, in social strata.

Cambridge in England, was produced in 1815. It is less spectacular than might be expected (on it, the Underworld looks merely decorative) for "the map that changed the world", as it was described by historian Simon Winchester in his 2001 book of the same name.[2] The mapmaker was William Smith, a self-educated geologist from London who – single-handedly and in great detail – located the coal, ores and other resources found beneath the surface of Great Britain, and thus laid the foundations both for great advances in geology and the life sciences, and for extensive mining. Fossil fuels became accessible on a huge scale.

In 1793, German anatomist Johann Christian Rosenmüller discovered large fossilized bones in caves in Bavaria in Southern Germany. He reasoned that they belonged to a species of ancient, extinct bear, different from any animal known at the time.[3] Such findings would invite new questions about prehistory and the extinction of animals. Early in the nineteenth century, Mary Anning of Lyme Regis, a town in West Dorset, England, discovered another set of fossils inviting similar questions.[4] An old seabed containing layers of bizarre fossils had been pushed above sea level along the Jurassic Coast, exposing the fossils to curious naturalists like Anning; she would explore the cliffs and hills with

Left: Mary Anning of Lyme Regis, Dorset, England, *c.*1812.

Below: Ichthyosaur, an important fossil discovered by Mary Anning in 1812.

Opposite: Cave paintings at Chauvet-Pont d'Arc in France, possibly indicating volcanic eruptions.

a rock hammer in hand. The fossils she uncovered would later prove to be 200 million years old. Among them were the remains of marine reptiles that came to be named ichthyosaurs. Rosenmüller and Anning's discoveries revealed that life had a very long history indeed.

Scottish polymath James Hutton, scientist and farmer, also deserves credit for drawing broad attention to the history of the Earth. In his book *Theory of the Earth*, sometimes regarded as the first treatise on modern geology, he skilfully drew upon geological observations to argue that the Earth was perpetually formed, establishing geologic and cosmic time – deep time.[5] For him, the key forces of geologic formation were erosion and sedimentation. These were slow processes, he reasoned, a godly design for keeping Earth eternally suitable for humans. While such a notion had Anthropocenic undertones,

presaging modern worries (with God in the role of Anthropos), it failed to capture sudden catastrophes. After all, Hutton was a farmer, observing his land in the slow motion of seasons and centuries. Yet in 1755, not long before he wrote his treatise, a terrible earthquake struck Lisbon, drawing attention to the occasional violence of geologic events at timescales more appropriate for the Anthropocene.

Hutton must have known that earthquakes and eruptions were important geologic agents. For a long time, volcanic craters had been seen as the potential entry into the history of life and Earth. In the book *Mundus Subterraneus* (*Underworld*), published in 1665, the German Jesuit priest and polymath Athanasius Kircher described the malevolent beings you would meet if you climbed down into a volcanic crater. Around a hundred years later, Alexander von Humboldt, a pioneer of

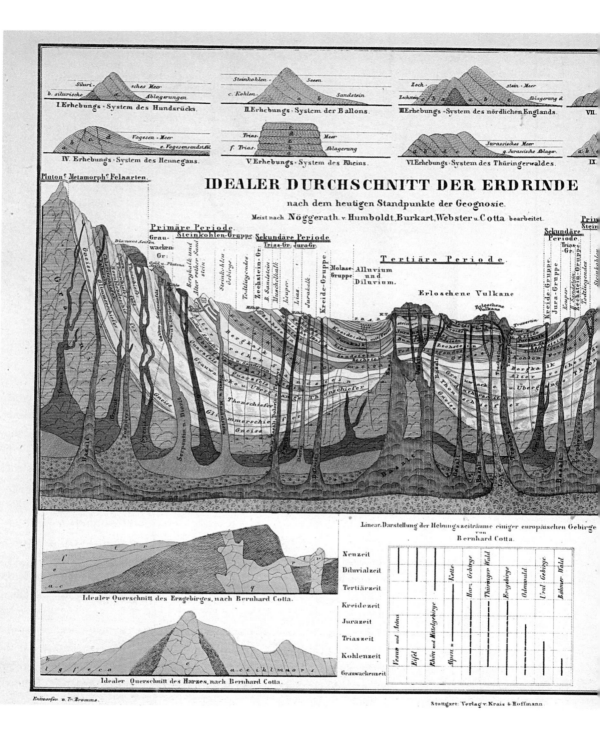

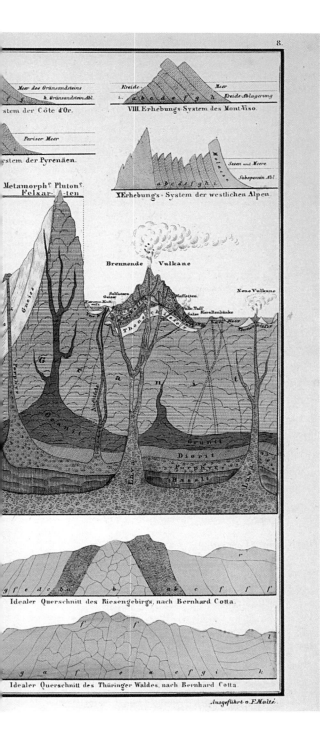

the earth sciences, argued that "a single volcanic furnace" was behind all the volcanic activity on the planet.[6] The causes of eruptions have been a persistent topic of fascination for millennia, particularly in regions with extensive geologic activity. Testimony to this are the 36,000-year-old paintings in the caves at Chauvet-Pont d'Arc in France. In these, possibly the oldest known eruption pictures, painted around the time that *Homo sapiens* was colonizing Europe, the paintings depict the volcano, its surroundings, and perhaps its mysterious underworld. It is known that about the time that the cave dwellers created their paintings, an eruption did indeed occur in a nearby volcano.

Now we know, partly thanks to the recently developed ArchaeoGLOBE project that assembles archaeological results from different parts of the world, that Anthropocenic signatures established themselves far deeper in time than Anning, Buffon, Hutton, von Humboldt and their contemporaries would probably have imagined. These signatures did so through the use of fire by reshaping the land and by driving species to extinction thousands of years ago.[7] The concept of deep time helps to get a perspective on the life-world of the planet, understanding previous warmings and coolings of the Earth, providing a reasonable sense of where we are heading, and revealing the gravity of the current warming we are experiencing. Humans alone have developed an understanding of deep time; this brings immense responsibilities to our hands.

Left: Alexander von Humboldt's "Idealer Durchschnitt der Erdrinde", 1851. From *Atlas zu Alex. v. Humboldt's Kosmos* (1851).

Early signs and warnings

Climate change is not a recent phenomenon in human history. During the Middle Ages, for instance, parts of the world experienced extended warm and cold periods, with associated famines, diseases and wars. Such shifts, of course, would be noted and recorded, partly because they affected everyday lives and future prospects, including food choices. Sometimes public accounts of radical change in climate, witnessed within a generation or two, were millennial or apocalyptic in style, predicting or unveiling catastrophic developments relating to inevitable end times. Renaissance scholars, for instance, speculated about the impact of deforestation, irrigation and the draining of wetlands on local climate, and on the broader ecological outlook. From what they knew of prehistory, however, the baseline assumption was probably that climate remained more or less constant, apart from exceptional periodic swings and seasonal fluctuations, and that human activities played no major role.

Scientific warnings about climate change and its potential environmental impact are probably older than most people think. The first discoveries in this field date from the early nineteenth century.[1] Change was not necessarily attributed to human-induced alterations of the environment, nor was it necessarily seen as a bad thing. Some of the pioneers working on evidence from the Palaeolithic era established the prior existence of ice ages, long periods with reduced planetary temperatures, while others identified the greenhouse effect, the heating of the atmosphere through the interaction of light energy emitted by the sun, the Earth's surface and atmospheric gases. While several contributing factors were suggested to account for climate change, including volcanic eruptions or fluctuations in solar radiation, over time, the

The Keeling Curve

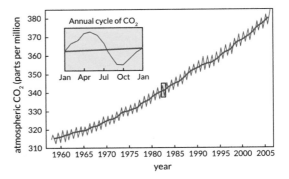

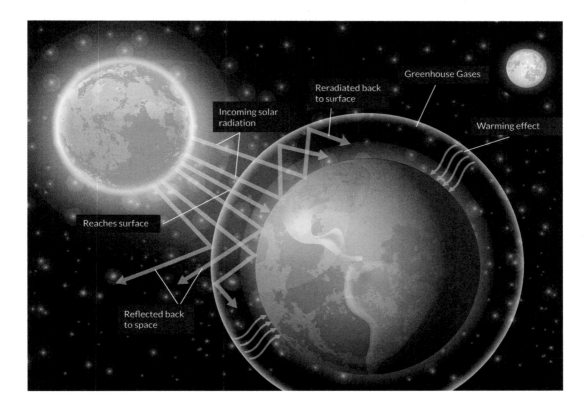

Incoming solar radiation

Reradiated back to surface

Greenhouse Gases

Warming effect

Reaches surface

Reflected back to space

emission of greenhouse gases came to be seen as the main source of change. Climate-change scholars have based their conclusions and predictions on several kinds of evidence: simple calculations, modelling, and the study of tree rings, ice cores and historical records. In 1896, Swedish physicist Svante Arrhenius calculated that declines in atmospheric levels of CO_2 (carbon dioxide), the primary greenhouse gas, would lower temperatures, resulting in ice ages. Conversely, doubling the CO_2 would increase temperatures by 5 to 6 degrees Celsius. Some scholars suggested that an increase in CO_2 would not necessarily result in warming, as the ocean might absorb the gas. In 1938, English engineer Guy Stewart Callendar presented evidence that

CO_2 level and temperature had been rising for decades, but the scientific community largely ignored his argument. In 1960, roughly half a century after Arrhenius's calculations, American chemist Charles David Keeling demonstrated that the level of CO_2 in the atmosphere was increasing and that things were, indeed, heating up. Based on his measurements on Mauna Loa, Hawaii, begun in 1958, his findings fuelled increasing worries about the damaging effects of human activities, in particular coal mining and the carbon industry.

Opposite: The Keeling Curve, showing atmospheric carbon dioxide concentrations at Mauna Loa, Hawaii, 1958 to 2019.

Above: The greenhouse effect of solar radiation on the Earth's surface caused by greenhouse gases.

In the 1890s, American astronomer Andrew Ellicott Douglass suggested that tree rings might reveal something about climatic time, noting that the rings were thinner in dry years. The value of tree rings was disputed for a time, but eventually established in the 1960s. Soon after that, ice cores provided another even more fruitful avenue into the climatic past. In the 1980s, international teams drilling into the glaciers of Greenland and Antarctica managed to show how climate had shifted from time to time, sometimes dramatically in relatively short timeframes that might be less than a human lifetime.[2] For many scientists, the dramatic moment that marked the "discovery" of abrupt climate change was a 1993 report on the analysis of Greenlandic ice cores.

As a whole range of historical observations became available, a pattern emerged. At the same time, computing technology advanced, allowing for the study of long time intervals and the modelling of climatic futures. In 1981,

American climatologist James Edward Hansen and his colleagues showed that human activities had affected global climate, predicting dramatic consequences in the near future:

There is a high probability of warming in the 1980s. Potential effects on climate in the 21st century include the creation of drought-prone regions in North America and central Asia as part of a shifting of climatic zones, erosion of the West Antarctic ice sheet with a consequent worldwide rise in sea level, and opening of the fabled Northwest Passage.[3]

Combining modelling and actual observations, scientists confirmed that greenhouse gases caused global heating. Hansen's public testimony to the United States Congress seven years later raised serious alarms domestically and internationally. A fairly broad scientific consensus seemed finally to be established.

Despite spectacular advances in research, the public perception of Anthropocenic climate change is not a linear progression from ignorance to understanding. Established environmental facts are sometimes challenged later on, because of lack of trust in science, the vested interests of ignorance, or some form of cultural and political backlash. Since the early twenty-first century, warnings of the detrimental impact of humans on climate have often been either ignored or systematically silenced, with serious consequences.

Sometimes the experts fail, jumping to wrong conclusions because of limited information or the inertia of their cosmologies. The most serious scientific error, in hindsight, is not the exaggeration of the problems associated with increasing levels of CO_2 in the atmosphere, but the opposite. For decades, scientists have been modest in their prognoses, assuming that change is a slow and distant prospect and that early worst-case scenarios were overblown. Representing a United Nations grouping of thousands of scientists, the Intergovernmental Panel on Climate Change of 1990 said that climate change would come at a stately pace and that the ice sheets of the Arctic and the Antarctic were stable. Economists predicted that the economic consequences would be small and stable. Now we know that things are happening much faster than expected and that the costs of responding have been grossly underestimated. In 2018, a United Nations report concluded that the world is headed for heating of 1.5 degrees Celsius over the next 80 years, but another more recent United Nations report now suggests that temperatures will rise by double this figure, with catastrophic economic and ecological damage.

Earth is now increasingly understood as a single, all-encompassing ecosystem, which includes humans and their activities. This is the so-called Gaia hypothesis, named for the ancient Greek personification of Mother Earth as the source of everything, from the heavens to humans. The word derives from the Greek root *gē* ("earth"), as does the modern term "geoscience".

When British chemist James Lovelock and American biologist Lynn Margulis first advanced the Gaia hypothesis in the 1960s, some called it pseudoscience, informed by romantic new-age thinking. Lovelock's book *Gaia: A New Look at Life on Earth*, published in 1979, urged people to see the Earth as a kind of living organism. It has had a dramatic impact.[4] While the Gaia hypothesis was received with scepticism at first, it now seems it has been fully vindicated, seen as a herald of the Anthropocene. People recognize the profound impact of organisms – including themselves – on the planet itself. Organic life is part of deep time, stretching deep into the planetary fabric.

Opposite: The exploration of climatic signatures in ice cores, Greenland.

Below: The key graph from historic 1981 *Science* article by James E. Hansen and colleagues, establishing human-caused global warming.

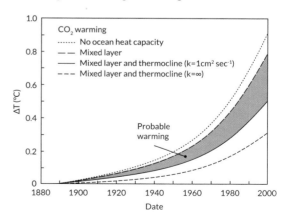

Fire and the Long Anthropocene

In 1954, American anthropologist Loren Eiseley argued that fire – which began as a manner of preparing the land, opening new space, keeping warm, cooking and perhaps managing enemies and spirits – was the "magic that opened the way for the supremacy of *Homo sapiens*"; humanity, he argued, was itself a kind of flame.[1] As the Anthropocene advanced, fire would lead to large-scale deforestation, transmuting the local environment. With growing demand for energy,

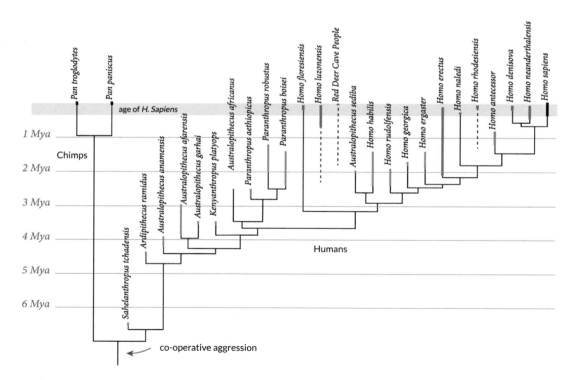

humans reached for new sources of fuel from the deep geological past, exhuming lithic landscapes. Fossil biomass – coal, oil and gas – proved to be essential to the new combustion technologies of the Industrial Revolution. Fire was the ingredient that set the wheels of the Anthropocene in motion, and what keeps them turning today.

Just as historians sometimes find it convenient to speak of a Long Eighteenth Century, to capture particular social transformations that eschew the normal centennial frame (for many British historians, it runs from the Glorious Revolution of 1688 to the Battle of Waterloo in 1814), we may find it useful to refer to the Long Anthropocene, spanning the entire period of human transformation of the planet. Not only have we significantly contributed to global heating and destruction during the last decades and centuries, but human activities have extensively changed

the surface of Earth and its biota for thousands of years. Several human species walked the Earth 300,000 years ago – including the Neanderthals, *Homo neanderthalensis*, who interbred with *Homo sapiens* and probably had an early Anthropocenic impact through the use of fire. The spread of *Homo sapiens*, who evolved 260,000 to 350,000 years ago in Southern Africa, may have resulted in the extinction of Neanderthals about 40,000 years ago.[2] Now it's just us.

It has now become evident that humans dominated Earth much earlier than previously thought – possibly up to a millennium earlier than often assumed. This broad picture has long remained hidden, as archaeologists have tended

Opposite: Graph demonstrating human evolution. Nick Longrich, *The Conversation*, 21 November, 2019.

Above: A Himba homestead in the Kunene Region of Namibia.

to focus on local sites and regional perspectives, failing to grasp overarching patterns. In 2018, a global collective of archaeologists came together to create the ArchaeoGLOBE project, a programme that drew on the results of around 250 experts to build a broader, more inclusive, comparative model of prehistoric land use. Their model showed that, as far back as 3,000 years ago, the planet had already been "largely transformed by hunter-gatherers, farmers, and pastoralists".[3] Intensive agriculture was the basis of the Mayan civilization in Mesoamerica and the Zhou dynasty in China from about 1,000 BCE. Pursuing the occupation and agricultural development of new land often meant burning it and shaping it for new modes of subsistence.

Early use of fire radically refashioned landscapes and ecosystems, destroying forests and possibly leading to some of the first extinctions driven by humans, notably the disappearance of megafauna. Sometimes, no doubt, the fires raged out of control, spreading dust particles over great distances, leaving telltale signatures in nearby glaciers for curious modern-day scholars to extract.

Fire has continued to play an important role in relations between humans and their environment. As environmental historian Stephen J. Pyne argues, the flames are now "descending as rapidly as their fuels are rising. They are burning through deep time" – a profound testament to the depth of human impact and the scale of the Anthropocene. Fire is not a material substance but a reaction, Pyne reminds us, an oddity in the company of the other three ancient elements: earth, water and air. While nature's fires used to be rare and patchy, often ignited by lightning, under human influence they gradually intensified to create a new era in which natural history, including climate, would become a subset of fire history: "A fire age was in the making".[4] Now, catastrophic fires are repeatedly afflicting the Amazon, California, Australia and elsewhere, leaving thousands of people without homes – and no doubt about the seriousness of global heating.

Given the "long" extension of the Anthropocene, some have argued, it makes sense to speak of the Capitalocene, the Pyrocene (the fire age that Pyne spoke of) or the Plantationocene rather than the Anthropocene. The deciding factor, it is often argued, is not *Homo sapiens* as such, but specific institutions and practices crafted by *parts* of humanity; namely, the Northern Hemisphere and the wealthy classes subjugating the rest of humanity to poverty and misery, impoverishing the entire planet in the process. These institutions and practices – not the all-inclusive, collective "us" – are perhaps the main causal factors that have set the planet on its current path of mass destruction, driving an industrial revolution, the birth of the carbon industry and the excessive production of CO_2 on an overcrowded, fuming planet.

Many significant Anthropocenic changes began with European colonialism in the fifteenth century. The discovery of the New World, the expansion of the nation state, plantation economies and the brutal institution of slavery all laid the foundations for the Industrial Revolution. This involved new kinds and degrees of inequality (such as those between enslaved people and slave holders, and between workers and owners of capital and machinery) as well as escalating ecological

Opposite: Forest fire in Altamira in Para state, Brazil, August 2019.
Overleaf: Fire storm in New South Wales, Australia, November 2019.

33

FIRE AND THE LONG ANTHROPOCENE

THE SAND CREEK MASSACRE

One of the ugliest colonial encounters was the Sand Creek Massacre of Cheyenne and Arapaho Indians in Colorado on 29 November, 1864, after years of tension. Seven hundred members of the Colorado Territory militia, led by US Army Colonel and Methodist preacher John Chivington, slaughtered 133 Indians, 105 of whom were women and children. A treaty between the Indians and the US had been signed in 1851, but a gold rush and other factors had pushed the US to renegotiate, as a result of which the Indians lost most of their land. Despite a new treaty in 1861, the massacre took place. At this time, the US had 29,000 miles of railways, requiring land and coal for the locomotives.

damage. Overexploitation of resources and radical alterations of local ecologies led to what would later be called "ecological imperialism".[5] Colonization inevitably implied confrontations of tribes, states and cultures that have extended into the present in one form or another.

The notion of the Plantationocene is probably the most serious competition to the Anthropocene, given its attention to lasting structural changes in human life, enslavement, global movement of goods and people, land use, and human–environmental relations.[6] Yet, again, only parts of plantation society bear responsibility for these changes: European rulers, traders, slave owners, shipping companies, doctors and priests – not the enslaved people who, by very definition, did not even own themselves.

Opposite: Cheyenne and Arapaho massacre at
Sand Creek, Colorado, November 29, 1864.

Top: The aftermath of the summer 2019 forest fires in
the Vila Nova Samuel region near Porto Velho, Brazil.

Above: Plantation slavery. Sugar cane
production in Barbados, c.1890.

FIRE AND THE LONG ANTHROPOCENE

SIX

The sad fate of disappearing species

The Long Anthropocene was driven by several developments: colonialism, the development of modern science, and a quest for natural resources, including mineral and fossil fuels. At the same time, it heralded threats to a number of species. It took a while, however, to see the effects of human activities.

Around the time that Svante Arrhenius and his colleagues were investigating the greenhouse effect, British zoologists including Alfred Newton were busily documenting escalating species extinction. Newton's pet project was the case of the great auk (*Pinguinis impennis*), a flightless bird and one of the most vulnerable species in the wake of European voyages to the New World. By the mid-nineteenth century, the great auk had become a signature species, a reminder of potential losses due to overexploitation. It was probably the first animal species knowingly

Opposite: Eldey, or the "Mealsack", as it was sometimes called. The "Underland" to the right was the last known breeding ground of great auks.

Above: *Pinguinus impennis* by John Gerrard Keulemans, *c.*1900.

Below: The great auk and its former distribution in the North Atlantic. The yellow zone shows the geographic distribution of the species while the sites marked with blue dots represent the sampling sites in a recent 2019 study by Jessica E. Thomas and her colleagues.

pushed to the edge of extinction by humans. As a result, it was extensively discussed in both academia and the public domain. The case of the great auk played an important role in early speculations on extinction, a pressing issue nowadays under threat of large-scale collapses.

Great auks were peaceful, social animals. In early summer, the parents would breed on islands and skerries (rocky outcrops) in small colonies among other bird species. They would lay one large egg, incubating for a few weeks on shifts. At five days old, the chicks would throw themselves into the sea. The skerries were not selected at random; here, breeding birds remained safe from animal predators for millennia, although both Neanderthals and early humans would hunt them for food, decoration and ritual.[1]

Early in the sixteenth century, Europeans had news of big colonies of great auks in Newfoundland. These colonies were hunted down within a century or so by seafarers from France and Portugal.[2] The vulnerable birds were herded together in great numbers and clubbed down. Once the fishermen had stacked their schooners with salted great auk meat, they sailed away; the supply would suffice for a whole crew on the way back home to Europe. The great auk, in other words, drove both European exploits in the New World and the Anthropocene, supporting the colonial regimes which in turn drove the great auk further down the road to extinction. Later on, when European collectors busily competed for rare birds, skins and eggs, the small colonies in Iceland and elsewhere in Europe began to collapse.

The last days of the great auk are fairly well known, thanks to the Icelandic expedition of egg collector John Wolley and his friend, the zoologist Alfred Newton, in the summer of 1858, and the *Gare Fowl Books* they wrote during and after the trip. Now kept at Cambridge University Library, the *Gare Fowl Books* (about 900 handwritten pages in five volumes) represent an underexploited source, offering a rare window into the perspectives of the last crews who hunted for the great auk under growing international pressure for museum specimens.[3] When Wolley and Newton left Victorian England to look for great auks in Iceland, one of their friends teased them, in a play on the animal's name, that this was a "genuinely *awkward* expedition".

Wolley and Newton hoped to see some great auks, unaware of the fact that the species had become extinct. They planned a trip to the island of Eldey, south of the Reykjanes Peninsula, in Iceland, the last known breeding ground of the bird, hoping to study its behaviour and habits and, perhaps, buy a stuffed bird. The weather, however, did not permit sailing to the island on an open rowing boat. To compensate for the lack of direct experience, they interviewed fishermen who had been on the last expedition.

During the last expedition, foreman Vilhjálmur Hákonarson had sent three of his crew to climb on the so-called "Underland" of Eldey, where the birds bred. One of them, Ketill Ketilsson, told Wolley the dramatic story, part of which has been widely circulated in the great auk literature:

Opposite left: Illustration titled *Catching Auks*, 1853.

Opposite right: John Wolley, British naturalist author of the *Gare-Fowl Books*.

Below: Engraving of the great auk after John James Audubon's illustration for *Birds of America* (1827).

Ketill, Sigurður [Ísleifsson], and Jón Brandsson landed. The former two ran together after one of the birds, but as they got near the edge, Ketill's head failed him and he stopped. Sigurður went on and seized the bird. My informant, torn as he was throughout the chase, went to look at the place from which the bird had started and there he saw an egg lying on the lava stab which he took up, it was cracked or broken. Ketill laid it down again where he had found it.[4]

Hákonarson sold the two birds to a Danish trader for the equivalent of £9, approximately £540 today. The story of the last expedition firmly established itself in popular culture. In 1863 Charles Kingsley wrote the novel *The Water-Babies*, an adventure for children featuring great auks on the brink of extinction.[5] The bird at the

Ootheca Wolleyana Tab.XX.

H. Grönvold. pinx.

Bale & Danielsson, Ltd. lith.

centre of the story recounts the history of its own species. *The Water-Babies* was highly popular for years, until it had to be set aside due to the racial prejudice implicit in the text, another product of the Anthropocene and plantation slavery.

Although enthusiastic bird watchers like Newton and Wolley helped to pave the way for an animal protection movement that later helped prevent the extinction of several species, they were not detached spectators, independent of the markets and hunting expeditions that destroyed rare species, including the great auk; on the contrary. They were direct participants, along with many others, among them European traders and Icelandic peasants, establishing a whole colony of stuffed birds in European and American museums.

Interestingly, a recent genetic study by Jessica E. Thomas (Bangor University and the University of Copenhagen) and her colleagues indicates that the great auk had not been under threat due to environmental change. By sequencing mitochondrial genomes from bone and tissue samples from 41 individuals across the species' historical geographic range in the North Atlantic, and reconstructing population structure and population dynamics, they were able to conclude that the species' genetic diversity was high and it was doing well until humans arrived in the Newfoundland in the early sixteenth century: "human hunting alone could have been sufficient to cause its extinction".[6] This was an Anthropocenic extinction.

The fate of the great auk, then, sensitized the public in many northern contexts to some of the damaging effects of what would be called the Anthropocene and to the need for environmental protection. Nowadays, with the advance of the Anthropocene, in the wake of rapidly escalating human impact on life on Earth, the case of the great auk, along with other "signature" narratives of extinction, poses new and pressing questions about the meaning of biodiversity and the loss of species. Homer reasoned centuries ago that birds were "winged words", carrying messages of one kind or another. What would the birds have to say to us about the Anthropocene, now that they are rapidly becoming extinct?

Opposite top: Vilhjálmur Hákonarson, leader of the last hunting expedition for the great auk.

Opposite: Early twentieth century illustration of great auk eggs.

Above: *The Water-Babies*, an adventure for children by Charles Kingsley, published in 1863.

2

Human Impact

....

The birth of extinction and endlings

The idea of species was practically non-existent until the seventeenth century. It was Englishman John Ray who first coined the term in around 1680. About half a century later, Carl von Linné outlined his system of biological classification in *Systema Naturæ*. Early on, most people assumed that a species was born once and persisted indefinitely; those that existed would not disappear and new ones would not arrive. Carl von Linné was only interested in existing species, and in the natural idyll he observed in rural Scandinavia during the eighteenth century, as if prehistory did not matter or was non-existent. "We will never believe," he boldly claimed, "that a species could totally vanish from the Earth."[1] Species of animals that were not visible, it was assumed, were simply lost or hiding. The French zoologist Georges Cuvier did raise the idea, amidst the turmoil of the French Revolution, that some species had disappeared for good, while a century earlier, the Comte de Buffon had detailed the evidence for prior extinctions, but their ideas were controversial.[2]

British biologists Charles Darwin and Alfred Wallace took over where Buffon and Cuvier had left off, demonstrating that life had a much longer history than previously believed and, moreover, that it was continually changing thanks to the fundamental role of natural selection. Each species had to have "evolved" from something

pre-existing. Although knowledge of extinction had prompted Darwin to speculate about biological variation and natural selection, he rarely mentioned "extinction" in his extensive work *On the Origin of Species* (1859). For him, extinction was inevitable and taken for granted; the competition of life forms necessarily pushed some forms aside, as one thing led to another in the continuous unfolding of life. The history of the Earth and of living things merged together over the expanse of "deep" time. Species had

Opposite left: British zoologist and ornithologist Alfred Newton.

Opposite right: Charles Darwin, the father of evolution.

Below: Magdalene College, Cambridge, where Alfred Newton worked.

THE BIRTH OF EXTINCTION AND ENDLINGS

evolved and disappeared in the distant past, long before humans arrived. Darwin and most of his contemporaries, in other words, were not interested in extinction in the present.

It was British ornithologist Alfred Newton, the first professor of zoology at Magdalene College in Cambridge, who arguably played a key role in establishing the modern concept of extinction as a topic for research and policy, paving the way for animal protection.[3] Like many other bird enthusiasts of the Victorian era, he was fascinated by the history of the flightless dodo (*Raphus cucullatus*) which became extinct on Mauritius in the Indian Ocean in the seventeenth century. European sailors would hunt the birds

and destroy their habitat. "Dodo" derives from a Portuguese word meaning "stupid", which says more about the hunters than the hunted. It was, however, the great auk which remained Newton's main focus for most of his life. He was keen to draw lessons from its sad fate. His concerns about the extinction of the great auk would, as discussed in the previous chapter, generate public support for the protection of species, drawing attention to the damaging impact of humans on natural habitats and the necessity of halting or reversing the damage. Although humans were part of the natural world along with the great auk, they were not obliged to remain silent and inactive. During the 1860s, Newton began organizing support for bird protection, arguing for, among other things, "closed seasons" (suspension of hunting during breeding seasons). More importantly, he played an important role in establishing the Association for the Protection of Birds and in facilitating the first national legislation in Britain on behalf of non-game animals: the 1869 Sea Birds Preservation Act.[4]

Early attitudes to bird protection varied from one country to another. In Germany, for instance, bird protection focused primarily on their usefulness for farming; species that were detrimental to agricultural land development enjoyed no protection and only a few people would miss them if they seemed to disappear, while species that kept harmful insects under control were respected and protected. In Britain, in contrast, rare species and those considered beautiful would be most carefully protected, irrespective of their practical usefulness. In this case, bird enthusiasts and naturalists led conservation efforts, not farmers.[5] Some birds were more equal than others, their fate dependent on a host of ethical, economic, contextual and aesthetic considerations.

Newton's key innovation with respect to the disappearance of species was to make a clear distinction between the slow and "natural" extinctions with which Darwin was concerned, and the "unnatural" extinction faced by the great auk and many other species at the middle of the nineteenth century, extinctions caused by humans. Thus, he was able to brush aside Darwin's fatalism, which suggested that nature was simply following its course, and hammer home the necessity of direct action to slow down or avert extinction wherever possible. In doing so, he opened a space for environmental expertise and the possibility of saving rare species on the brink

of extinction. Biology and natural science, he reasoned, would play central roles in the future, beyond – if not above, in Newton's view – the reach of laypersons and politicians. Extinction through human intervention, he insisted, was both unnatural and avoidable.

Although Newton was an inveterate collector of specimens, classifying and recording them in the fashion of the Victorian age, he went much further, treading new and important ground.[6] It seems ironic that the man who put extinction on the modern agenda remains more or less unknown outside a narrow circle of bird enthusiasts. His approach was controversial, partly because it challenged dominant ideas at the time about detachment and neutrality. However, with the new legal frameworks for the protection of species, which Newton himself helped pioneer, his beliefs eventually began to resonate and gain respect, and today many of his concerns are reflected in the development of concepts such as "biological diversity". Similar ideas about extinction and the need for protection were proposed in the United States, around the same time, by environmentalist George Perkins Marsh, who warned against the damaging effects of human activities in his book *Man and Nature*.[7]

Collectors and naturalists of Marsh and Newton's generation were not innocent participants. Often, they had collected birds and eggs of rare species without asking anyone for permission, sometimes knowing that the species in question was in imminent danger of extinction. Naturally, the understanding that a species was becoming exceptionally rare affected the market prices that specimens could command, often with

Opposite: Engraving of dodo from *Cassell's Natural History* (1896).

THE BIRTH OF EXTINCTION AND ENDLINGS

THE LAST SNAIL

The focus on the irreplaceable, final organism – whose death marks the extinction of a species – continues to this day, steadily escalating with public awareness of mass extinction. In January 2019, the *New York Times* announced the death of the last land snail (*Achatinella apexfulva*) in Hawaii. The snail's caretakers called him George, after the last Pinta Island tortoise (*Chelonoidis abingdonii*) of the Galápagos, known as Lonesome George, who died in 2012.[9]

Opposite: British bird eggs from *The Boy's Own Annual* (1896).

Above: Lonesome George, a male Pinta Island tortoise and the last known individual of the species.

Right: The last land snail (*Achatinella apexfulva*), Hawaii 2019.

the species' very rarity driving it to extinction. Such "last specimens" often carry extreme price tags; ironically, having been hunted precisely so that the museums could display them in their cabinets, they become too valuable for public display. Priceless becomes practically useless.

Significantly, the growing need to personify the last organism of its kind has generated new terminology. In the 1990s, American physician Robert Webster sent a letter to the journal *Nature*, launching the notion of "endling", defined as the last person, animal or other individual in a lineage. The idea came up because of patients who were dying and thought of themselves as the last of their family line. "Endling" seems to have stuck, outlasting competitors such as "ender", "terminarch", "lastoline" and "relict". As environmental historian Dolly Jørgensen has shown, "endling" has seeped into popular culture, becoming the focus of exhibits, philosophical writings, a musical composition and a ballet.[8] Repeatedly, reported endlings are mysteriously survived by others of their kind, a fact that is also highlighted in popular culture.

....

Enter the Industrial Revolution

Early agriculture and commerce facilitated human experimentation with machines, driven by human or animal labour, wood, wind or water. Leonardo da Vinci and Athanasius Kircher developed futuristic mechanical designs. Thus, in his work *Musurgia Universalis*, published in two volumes in Rome in 1650, Kircher sketched a complex hydraulic organ with cylinders and automata, powered by running water. Genuine industrialization demanded stable energy supplies on a new scale. Plantation slavery and, above all, the burning of fossil hydrocarbons, set in motion industrial production and mass transport, reorganizing human–environmental relations and accelerating the Anthropocene.

Since 1800, humans have consumed non-renewable energy – coal, oil and gas – on an unprecedented scale, channelling it into machines, electricity, artefacts, science and art. This is the essence of the Industrial Revolution. Soon it would refashion the landscape, constructing canals and railways, first in the heart of the revolution in Britain, and later throughout growing industrial powers, including South Korea, Canada and Brazil. The steam locomotive and ship carried their own energy supplies, vastly expanding networks of production, transport and commerce – and, of course, environmental damage and social inequality.

Industrial towns demanded disruption of peasant agriculture, assembling an impoverished working class for the expanding workplaces called "factories". In 1835, Alexis de Tocqueville wrote of Manchester, in England:

Thirty or forty factories rise on the tops of the hills ... The wretched dwellings of the poor are scattered haphazard around them. Round them stretches land uncultivated but without the charm of rustic nature ... You will hear the noise of furnaces, the whistle of steam. These vast structures keep air and light out of the human habitations which they dominate; they envelope them in perpetual fog; here is the slave, there the master; there is the wealth of some, here the poverty of most.[1]

Opposite top: Hydraulic organ, sketch from Athanasius Kircher's *Musurgia Universalis*, published in Rome in 1650.

Opposite: Steam engines driving English textile factories during the eighteenth century.

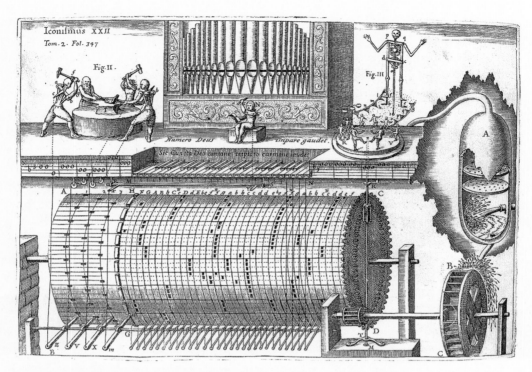

ENTER THE INDUSTRIAL REVOLUTION

The harsh conditions of life in the early years of the Industrial Revolution were captured in both novels, notably those of Charles Dickens, and social history, including Friedrich Engels's classic account of Salford, near Manchester, in *The Condition of the Working Class in England*, published in 1845. Dickens, who worked in a bottle factory at the age of 12 when his father became bankrupt, was sensitized to the brutality, dishonesty, and injustice of industrial society. In *David Copperfield*, he describes the gruelling process of paste-blacking bottles:

I know that ... certain men and boys were employed to examine them against the light, and reject those that were flawed, and to rinse and wash them. When the empty bottles ran short, there were labels to be pasted on full ones, or corks to be fitted to them, or seals to be put upon the corks, or finished bottles to be packed in casks.[2]

The Industrial Revolution had a major impact on the planet. Mechanized agriculture and intensive irrigation resulted in increased concentration of CO_2 and other greenhouse gases in the atmosphere, massive sediment loss, chemical pollution, altered water systems and species loss. Enclosures and privatization of land meant opening new spaces for large-scale agriculture, with the removal of hedges that were beneficial for many species, notably birds. The expansion of empire extended the domain of the West to almost every corner of the world, colonizing the territories of indigenous groups, including hunter-gatherers and pastoralists with the notable exception of the Arctic, which remained

ENTER THE INDUSTRIAL REVOLUTION

difficult to access for western explorers until the twentieth century. Now, interestingly, the Arctic is often seen as the canary bird of global heating, signalling the dangers of melting sea ice and the release of methane, a greenhouse gas, which had been trapped for millennia in permafrost. During the Industrial Revolution and the rise of modern science, people thought of the Earth as inanimate. Life and Earth became two separate worlds, as the latter became a resource, a platform for the world above – the human world. Scientists strove to view the Earth from a distance (even more so than during the Renaissance, during which artists' representations of perspective had illustrated and even perpetuated the widening gulf between the human race and the natural world), metaphorically jostling to establish the impartiality of the human observer.[3] Objectivity and detachment became scientific catchwords.

The land and the oceans became infinite reserves of resources and energy for human exploitation.

Such a separation of the human world and natural resources continued into the late twentieth century. This is graphically illustrated in environmental modelling, for example, in the 1988 natural science diagram of the "Earth's system" – the so-called "Bretherton diagram", which reduced the scope of "human activities" (and human sciences) to a small and unspecified box within a mechanical system of the Earth. While more recent versions of the Earth system tend to be more nuanced and to allocate more space for human activities, they usually fail to capture the depth and essence of the Anthropocene.

Opposite: Fuming chimneys in Stoke-on-Trent, UK, 1938.

Below: A simplified version of the Bretherton diagram of scientific modelling, 1988.

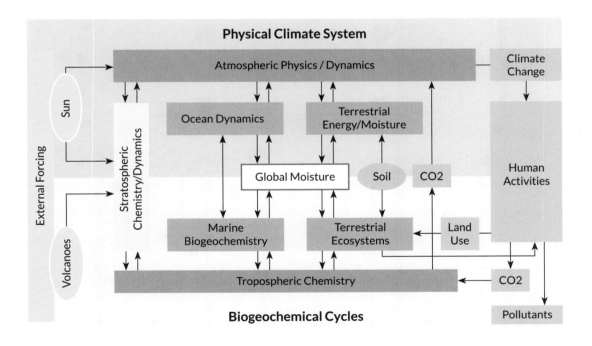

ENTER THE INDUSTRIAL REVOLUTION

It has become increasingly complex to model human–environmental systems in a realistic and accessible manner; Anthropocenic loops are becoming ever more twisted and convoluted.

It has been argued, possibly most prominently by Klaus Schwab, executive chairman of the World Economic Forum, that there have been four industrial revolutions.[4] Each revolution is characterized by particular innovations and sources of energy: 1) the original one was driven by the use of coal and steam power for the purpose of mechanizing production, 2) another one was

Below: Anthropocenic loops in the fourth industrial revolution.

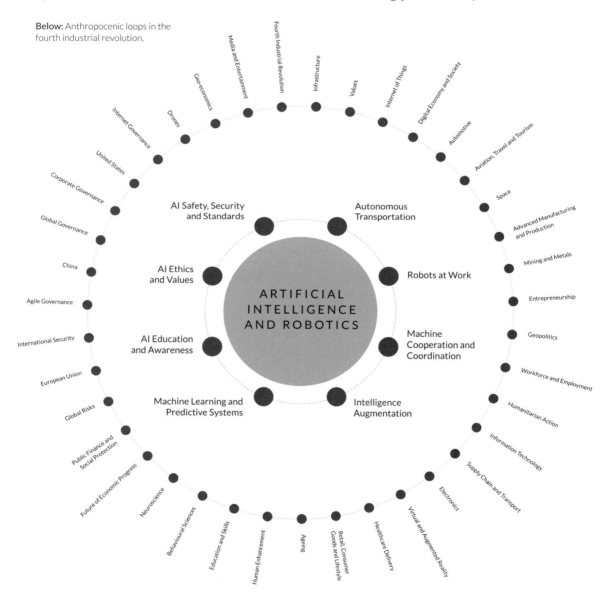

characterized by the use of electric power to enable mass production, 3) a subsequent one was highly dependent on electronics and information technology for automating production, and, finally, 4) the digital revolution is developing robotics and artificial intelligence. Each phase escalates anthropogenic impacts and blurs boundaries that have previously been seen as immutable – the demarcations between mental, animal, biological and technological.

In the future, most of humanity will probably live in megacities far bigger than we have ever seen. Unlike most of their predecessors, future cities will have to be carefully planned in advance, not expanded and designed ad hoc in the process of growth. Providing homes and shelter for environmental refugees and an expanding human population on a damaged, limited planet, these cities will need to draw upon the innovations of the industrial revolutions – modern architecture, electronic technology,

sensors, algorithms and digital machines – but they will also have to facilitate bicycles, green living and sustainable gardening, abandoning fossil fuels and heavy carbon footprints.[5] In this way, our future way of life is dependent upon those very industrial revolutions that have brought about our destruction.

Above top: Megacities would be home to 2,000 people per km^2.

Above: A rendering of a Lagos of the future, with a city design that works with rising sea levels.

N I N E

··········

The atomic age

During the second half of the twentieth century, human impact on the planet and life itself escalated to an unprecedented degree, partly in the shadow of the Cold War. The idea of the Anthropocene was in the air, although it didn't yet have a name, and there was a general sense of progress, not decline. Biology made great leaps, exploring the universe of cells and the nature of inheritance. Rosalind Franklin photographed DNA (1952), James Watson and Francis Crick discovered the double helix (1953), and the end of the century

produced a first draft of the human genome. Other explorations expanded our horizons into space and infinity.[1] Among the critical events taking place on this front were the launching of the satellite Sputnik 1 (1957), cosmonaut Yuri Gagarin reaching orbit (1961), landing humans on the moon (1969) and, later, the establishment of the International Space Station (1998).

Below: Design of a nuclear power plant.

Opposite: The first plan of a nuclear reactor, drafted by Enrico Fermi.

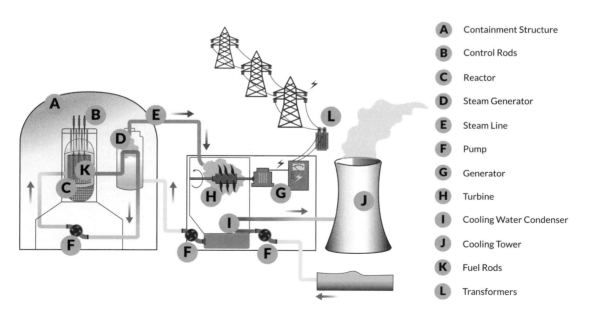

A Containment Structure
B Control Rods
C Reactor
D Steam Generator
E Steam Line
F Pump
G Generator
H Turbine
I Cooling Water Condenser
J Cooling Tower
K Fuel Rods
L Transformers

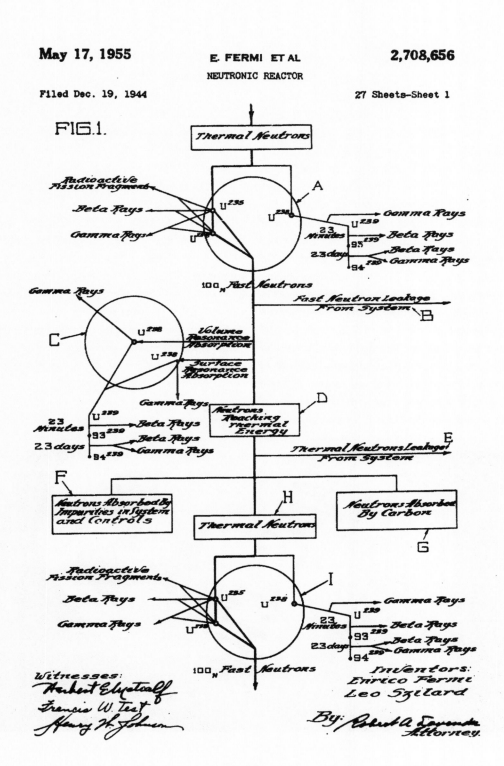

All of these developments drew heavily upon physics and digital technology. The most significant Anthropocenic impact, however, involved nuclear energy, the harnessing of nature's power on a scale vastly beyond anything seen before in human history. This was done either for peaceful purposes, solving the energy needs of an expanding industrialized population, or as part of a military game running out of control, with the potential to shatter the Earth and threaten all life. For many geoscientists, the age of the Anthropocene began with nuclear energy.[2] The first nuclear reactor for the purpose of producing a chain reaction was installed by Enrico Fermi in the swimming pool of the University of Chicago.[3] On 2 December 1942, in the middle of the Second World War, Fermi achieved the first controlled fission of atomic nuclei in a chain reaction. Two and a half years later, at 5:29 a.m. on 16 July, 1945, the first nuclear device was detonated. This was later known as the Trinity explosion, after a

THE BOMB

Three weeks after the Trinity test, the United States detonated a nuclear weapon over Hiroshima, Japan, and another one three days later at Nagasaki. The impact of the two bombings was shocking, killing between 129,000 and 226,000 people, mostly civilians. Some died instantly as a result of the shock and heat of the blast, others much later from injuries and radiation. The mushroom clouds produced by the detonations became icons of the nuclear age and its alarming threats. Concerns over nuclear waste, radiation and potential disasters sparked an international anti-nuclear movement, partly setting the stage for current apocalyptic worries about the Anthropocene.

Below: Nuclear waste.

Opposite: The Trinity explosion in the desert of New Mexico, 16 July, 1945.

THE ATOMIC AGE

code name assigned by J. Robert Oppenheimer, Director of the Los Alamos Laboratory. The same day, Oppenheimer remarked: "We knew the world would not be the same".[4] He was probably more concerned with the geopolitics of empires than what we would now call Anthropocenic impact.

The worst disasters of the nuclear age have been the incidents at the power plants at Chernobyl, Ukraine, on 26 April 1986, and Fukushima, Japan, on 11 March 2011. As a result of the explosion of Unit Four of the Chernobyl reactor, about 600,000 cleanup workers were affected and five million people were resettled from contaminated territories. Even though the Fukushima disaster was triggered by an earthquake and tsunami, neither it nor Chernobyl can be considered "natural disasters", as both resulted primarily from human action, flawed design and human error.

A vital source on the Chernobyl disaster is Belarusian Nobel Laureate Svetlana Alexievich's *Voices from Chernobyl: The Oral History of a Nuclear Disaster* (originally published in Russian in 1997). At the time of the disaster, Alexievich was a journalist in Minsk, about 400 kilometres from the Chernobyl plant. Based on hundreds of interviews with witnesses from the scene, her work highlights the human suffering of nuclear disaster.[5] In 2019, the Chernobyl incident re-entered public consciousness with acclaimed television drama *Chernobyl*, based on Alexievich's book. While some critics characterized the series as "disaster porn", it certainly made the disastrous consequences of the incident real for a new generation.

One important account of Chernobyl that has received much less attention than it deserves is American anthropologist Adriana Petryna's *Life Exposed: Biological Citizens after Chernobyl.*[6] Petryna's analysis draws attention to the failures to learn from the mistakes that caused Chernobyl and the attempts to silence or dismiss evidence. While the Chernobyl case should have been treated as a definitive "laboratory" for improving safety measures around nuclear power, twenty-five years later "the unlikely" happened again, in Fukushima.

It is estimated that about 400 times more radioactive material was released from Chernobyl than as a result of the atomic bombing of Hiroshima and Nagasaki. Petryna's account exposes lamentable failures to explore what happened both at the time of the disaster and in its aftermath. She explores the issue of what counts as truth, and concludes on a general note about the nature of evidence, highlighting that we need to attend to "the laws governing how biocomposites of whatever sort decay – to rules that determine how they are put away ... into the lithic world and, ultimately, cast off into geologic deep time", pointing out that fossils are "never just pure samples"[7]. The artefacts of deep time are not in stasis, but are instead subject to outside influence. This is a point that seems pertinent when establishing both recent and prehistoric Anthropocenic impacts on the planet.

Collectively, the history of military and nuclear impacts, from Hiroshima, Nagasaki and post-war testing, to the disasters of Chernobyl and Fukushima, draws attention to the perils of the nuclear age and their persistent threat in a world that is hungry for energy and plagued by arms races and instability. The alternative view still persists that nuclear energy harnessed safely could provide a greener option.

Given the successes and failures of the atomic age, some people have wondered if the future of humanity lies in outer space. While space poses particular problems for a species thoroughly adapted to terrestrial life, these might at least partly be suppressed by means of nuclear energy and other technological novelties. British physicist Stephen Hawking suggested that humans might be forced to leave Earth for environmental reasons. The mere recognition that settlements in outer space *are* a possibility underlines that space is just as "natural" for humans as Earth. Humans have made rapid progress in space exploration over recent decades, and progress that seems likely to continue.[8] However, relocating to outer space would be immensely costly and only for the privileged few, and is never likely to be an inclusive option. Thinking of the inequality generated through colonialism, Mahatma Gandhi saw an ironic twist to the issue of the human occupation of outer space. Asked whether independent India would follow the British pattern of "development", he replied: "It took Britain half the resources of the planet to achieve this prosperity. How many planets will a country like India require?". How many planets will the world require to clean up the mess of the Anthropocene?

Previous pages: Clearing the Chernobyl power plant in Ukraine after the disaster of 1986.

Opposite: Still from HBO's *Chernobyl* miniseries, 2019.

Below: Artist's conception of a never-realized spacecraft that was intended to be powered by a fission reactor.

Draining wetlands

Representing a substantial part of the Earth's land surface (about 6 per cent), wetlands (variously known as bogs, fens, marshes, or swamps) occur on every continent except Antarctica, and in every zone and biome. There are two wetland regions with areas in excess of a million square kilometres, the Amazon River Basin and the West Siberian Plain, and seven more that are between 100 and 400,000 km². Despite their ubiquity and scale, wetlands have typically remained at the margins of social discourse. Perhaps as a consequence, humans reduced global wetland areas by 50 per cent in only a century, and by nearly 90 per cent since the start of the Industrial Revolution, all without much concern or debate.[1] Towards the end of the twentieth century, however, wetland areas began to be recognized as constituting some of the most ecologically valuable areas on Earth for processing waste and for capturing greenhouse gases.

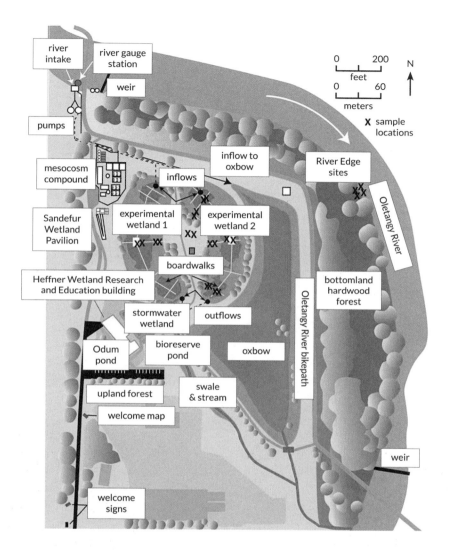

Map labels:
- river intake
- river gauge station
- weir
- pumps
- mesocosm compound
- inflows
- inflow to oxbow
- River Edge sites
- Oletangy River
- Sandefur Wetland Pavilion
- experimental wetland 1
- experimental wetland 2
- boardwalks
- Heffner Wetland Research and Education building
- stormwater wetland
- outflows
- bottomland hardwood forest
- Oletangy River bikepath
- Odum pond
- bioreserve pond
- oxbow
- upland forest
- swale & stream
- welcome map
- weir
- welcome signs

Scale:
0 — 200 feet
0 — 60 meters

N

X sample locations

To artists such as Dante, Milton and Ibsen, wetlands represented an infernal domain, where disease and nefarious acts were rampant. Dante said that wetlands encircled four of the innermost circles of hell, where heretics and those who deliberately lie and cheat are tortured until the day of doom. Set in the Fens of England, near Cambridge, Graham Swift's 1983 novel *Waterland* offers a modern version of Dante's hellscape. At the same time, it presents a series of intriguing observations of landscape and water: "The great, flat monotony of reality; the wide, empty space of reality. Melancholia and self-murder are not unknown in the Fens".[2]

Opposite: The Fens of England.

Above: Diagram of experimental wetlands demonstrating decreases in the emission of greenhouse gases.

DRAINING WETLANDS

Missouri River

Arkansas River

Ohio River

Tennessee River

Red River

Atchafalaya

Gulf of Mexico Hypoxia

0   500
Kilometres

Opposite: Hurricane Katrina flooding
New Orleans, Louisiana, 2005.

Above: Mississippi-Ohio-Missouri River Basin
in USA showing areas of land drainage.

Mississippi River basin (MRB)
■ Major nitrate sources in MRB
■ General extent of hypoxia in Gulf of Mexico
— Mississippi River basin boundary
● 8000 hectares of drained land in MRB

Wetlands have also, however, been seen as holy ground, symbols of life and renewal. The proponent of this view was the philosopher and environmentalist Henry David Thoreau, who emphasized that our ideas about nature always reflect what lies within: "It is in vain to dream of a wilderness distant from ourselves, there is none such. It is the bog in our brain and bowels, the primitive vigour of Nature in us that inspires that dream".[3]

Recognition of the importance of wetlands is reflected in an international convention, signed in Ramsar, Iran in 1971: the Convention on Wetlands of International Importance. The Ramsar Convention provides for action and international cooperation that contributes to the protection and intelligent utilization of wetlands. Currently, 171 countries have signed the convention. A key aspect of the Ramsar agreement was an inventory of wetlands, which led to the identification of over 2,300 wetland areas around the world, all considered important in an international context.

International studies to approximate the annual value of wetlands, in the context of their natural capital and the ecosystem services they provide, have assigned them a price tag of $12,790 trillion, equivalent to a third of the total value assigned to the entire world.[4] Dubious valuations aside, a metaphor frequently used with respect to wetlands is that of "biological supermarkets", on the grounds

that they are characterized by extreme biodiversity and substantial biomass. The assertion is also often made that wetlands are "biological machines" or even the "kidneys of the environment", a reference to ecological services they provide, purifying waste from humans and other organisms.[5] It is now more widely appreciated that draining wetlands damages the environmental kidneys and causes the biological machines to break down. It can also exacerbate flooding, with catastrophic consequences. Thus, the flooding of New Orleans in the wake of Hurricane Katrina in 2005 was partly the result of drainage projects. Ecologists have predicted that such catastrophes are likely to recur.[6] Any discussion of wetlands needs to avoid narrow definitions of ecosystems, and to take into account their dynamic nature, as well as the mutual interdependence of human activities and the communities and environments in which they are embedded.[7]

In the Enlightenment era, wetlands underwent more radical changes at the hands of humans than had previously been known. Innovations in agriculture and machinery such as excavators, tractors and ground-levelling equipment led to large-scale drainage of wetlands. Technology thus made it possible to manage and order wetlands to meet the demands of the economy. The mindset of this era regarded the marsh as counterproductive and unappealing.

In the eighteenth and nineteenth centuries, marshes and wetlands constituted obstacles to progress. This attitude reached its climax in a series of grand engineering schemes. One example is a large irrigation project implemented in southern Iceland during the first half of the twentieth century. In 1914, the Icelandic authorities financed a Danish design to facilitate flexible water management by individual farms, and to increase overall agricultural productivity. The project was extremely expensive, but the results were disappointing. Ironically, when the project was "completed", it turned out to be more or less obsolete, due to other innovations in agriculture.[8] This incident seems to illustrate the counterproductivity of the attempt to decimate wetlands, highlighting that to live in harmony with these habitats is mutually beneficial to the human race and the wetlands themselves.

In more recent times, a strong social movement has developed that advocates the reclamation of wetlands in many parts of the world. Many regions that were heavily drained have seen the rebirth of wetlands, with renewed vegetation and bird colonies. One of the success stories is that of Ukraine's section of the Danube Delta, thanks to the project Rewilding Europe.[9] Here, shoals of fish have returned and otters and birds have established new territories in the wake of the removal of 11 earth dams built in the 1970s; the wetland is recovering amazingly fast. The reclamation of wetlands is now seen as one of the major actions necessary to slow or halt global heating, given the capacity of wetlands to capture greenhouse gases. In 2019, the Intergovernmental Panel on Climate Change, pointed out that reclamation of wetlands works fairly quickly for sequestering carbon, with "immediate impact", comparing favourably to measures that take decades to deliver measurable results.[10]

Opposite: Wetland drained by a ditch.

Below: The grand engineering scheme of draining wetlands in Southwest Iceland, 1914.

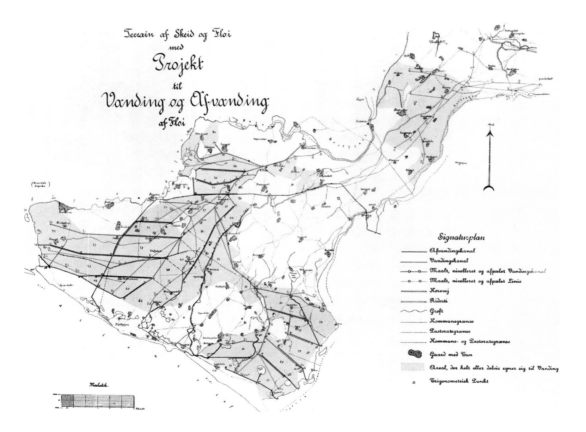

DRAINING WETLANDS

....

Plastics:
Broth, soup and islands

The concept of plastic has a long history. In his treatise *Meteorology*, written around 350 BCE, Aristotle identified two kinds of "impressibles": inanimate things that can be moulded by the human hand. Some impressibles (like wool), he suggested, were "squeezable" but did not retain their shape, while others (like wax) were *plastikus* (from the Greek *plassein*, "to mould"), retaining their form in the process of making.[1] The English adjective "plastic" has been around since the end of the sixteenth century, whereas the noun "plasticity", the ability to be moulded (as in references to the plastic mind of the child or the plastic neurology of the brain), seems to appear a little later, in the late eighteenth century. What is now called "plastic surgery" is mentioned in ancient Egyptian and Indian texts, but it was only during the second half of the twentieth century, with the advance of biotechnology and genomics, that the idea of plasticity was extended to life itself, to the refashioning of living cells.

There are good reasons for the popularity of plastics and their continued development in many forms: they are cheap to produce (depending on oil prices); their production generates lower greenhouse gas emissions than, say, paper; they are strong and waterproof; they help to reduce food waste; they can take various forms (especially with the aid of three-dimensional printing), and, most spectacularly, some types of plastic can be melted and reshaped time and time again. Indeed, plastics are partly responsible for growth in recent decades. The environmental costs, on the other hand, are enormous.

Plastic in the sense of synthetic material derived from petrochemicals – an ever-present reality in the modern world – was originally developed a century ago. Since then, it has found its way into all kinds of things. Images of vast rafts of plastic floating on the ocean have alerted many people to the growing threat of plastic products thrown overboard from sailing vessels or dumped into the ocean via sewers and rivers, accumulating in particular spots as a result of major ocean currents, called gyres. In 1997, marine captain Charles Moore noticed massive collections of bottles floating in the middle of the Pacific Ocean. He called the phenomenon "plastic islands" and

Opposite: A plastic island off the coast of South Africa, 2019.

73
PLASTICS: BROTH, SOUP AND ISLANDS

PLASTICS: BROTH, SOUP AND ISLANDS

helped to put the issue on the international environmental agenda.

Islands is a misleading term, as the plastic that Moore observed floats temporarily on or near the surface, eventually sinking to the ocean floor or hovering between the surface and the bottom. "Plastic soup" has been suggested as a more suitable term, highlighting the mixing of plastic and other material between the ocean floor and the surface. Still others recommend "plastic broth" to emphasize the formation of microplastic and nanoplastic, tiny fragments that go unnoticed as they enter the food chain, both marine and terrestrial. As they become parts of tissues and organs, they can never be removed.[2] Plastic "ghost nets" that get lost during fishing are not only a threat to travellers. By fouling boat propellers and floating freely in the ocean for decades, they affect fish, turtles, dolphins and whales, injuring and killing them. Small plastic objects can be mistaken by sea birds for food, with lethal consequences.

Perhaps the most dangerous aspect of plastic pollution is the phenomenon of nano- and microparticles infecting the food chain at practically every level. It is estimated that eight million tons of plastic enters the ocean every year while only a fraction of it, perhaps 1 per cent, is found on beaches or in oceanic "plastic islands". Parts of the missing plastic ends up on the ocean floor and the rest breaks down into tiny particles that are difficult to detect.

Snowflakes, it now emerges, are no longer the idyllic image of the purity of nature, for they are tainted by travelling microplastic.[3] The tiny fragments are carried by the wind over long distances, to the Swiss Alps and the Arctic. This development, still poorly understood, implies that microplastic circulates in the atmosphere, as a significant component of air pollution. Microplastic also rains down on city dwellers; London has the highest levels recorded so far.[4] This is the air we, and other creatures, breathe.

Originally seen as a sign of progress, plastic is now officially a nuisance. Sir David Attenborough, whose second *Blue Planet* TV series (2017) helped to raise the profile of plastic pollution in the ocean, recently predicted that polluting the planet would soon provoke as much abhorrence as human slavery.[5]

Opposite: A sea turtle caught in ghost nets.
Below: Kamilo Beach, Hawaii, is now known as Plastic Beach.

PLASTIC CULTURE

Plastic has, since its development in the twentieth century, become an ubiquitous part of popular culture and everyday language. Andy Warhol captured the essence: "I love LA. I love Hollywood. They're beautiful. Everybody's plastic – but I love plastic. I want to be plastic". Sculptor Peter Ganine crafted a yellow duck in the 1940s which he patented and reproduced as a floating toy. The "rubber" duck, a stylized duck with a flat base, made of rubber-like material such as vinyl plastic, achieved iconic status, and was often symbolically linked to bathing. Over fifty million copies were sold. Jeff Koons' seminal sculpture *Rabbit* (1986), made of steel, is one of the most iconic works of the twentieth century. In 2019, it sold for $91,075,000, the highest price ever paid for a work of a living artist. When it was first shown at a gallery in New York, the art critic for *The New York Times* described it as an "oversize rabbit, with carrot, once made of inflatable plastic".[6] In 2007, a monumental blow-up version featured in the Macy's Thanksgiving Day Parade in New York. Perhaps the banal, shining and steely imitation of a plastic rabbit is the ultimate symbol of the Anthropocene.

A number of visual artists have focused on the more sinister side of plastic. Florentijn Hofman, for instance, took the iconic yellow duck, which has travelled so many bathtubs and high seas, and blew it up to colossal scale; one of his gigantic rubber sculptures was 26 metres wide, 20 metres long and 32 metres high. Hofman has deployed them as temporary installations in the harbours of several major cities. Benjamin Von Wong's series of photographs, *Mermaids Hate Plastic*[7], present a liminal being combining human and animal features, enmeshed in a sea of 10,000 plastic bottles borrowed from a waste management centre. In the photographic series *7 Days of Garbage*, photographer Gregg Segal invites friends and family to immerse themselves in a week's worth of their own garbage – often a shocking experience.

Below: Rubber Duck by Florentijn Hofman, on display in Kaohsiung, Taiwan, 2013.

Opposite: Jeff Koons' *Rabbit*, stainless steel, 1986.

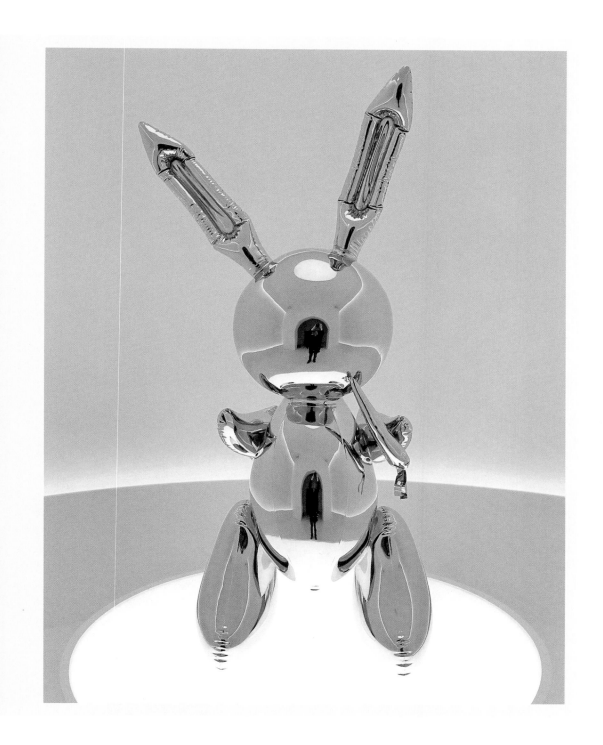

PLASTICS: BROTH, SOUP AND ISLANDS

One of the clearest signs of Anthropocenic impact on the planet is the existence of so-called plastiglomerates. These are aggregations that contain mixtures of naturally occurring materials (sedimentary grains and shells, for instance) glued together by hardened molten plastic, as a result of campfire burning. Such stones were first discovered on Kamilo Beach, Hawaii.[8] They are genuine geosocial formations and many of them will remain visible indefinitely to any geoscientists looking for clear signs of the impact of humans in the current age.

In a remarkably short time, plastics have made their mark on the planet as one of the key characteristics, if not *the* spike or signature geologic layer, of the Anthropocene. Plastics pose possibly insurmountable, environmental problems; once created, they refuse to disappear. One of the major issues on the current environmental agenda, locally and globally, is to minimize the damage of plastics, slowing down production, regulating their use and waste, and clearing the rubbish where possible. While individual efforts to contain plastic do help – and public attitudes do seem to be changing – governmental regulations, including the banning of single-use plastics, have a far greater impact in the attempt to stem the tide.

Opposite: *7 Days of Garbage* by photographer Gregg Segal, 2014.

Above: Plastiglomerate – geosocial rock – from Kamilo Beach, Hawaii.

Overleaf: Photograph by Benjamin Von Wong from his series *Mermaids Hate Plastic*, 2016.

TWELVE

Superheating

To see where we are in the midst of the Anthropocene, a historical perspective is essential. Donald Worster, one of the pioneers of environmental history, argued: "we ... have two histories to write, that of our own country and that of 'planet Earth'", and "when that larger planetary history gets fully written, it will surely have at its core the evolving relationship between humans and the natural world".[1] Such a reminder was timely in the 1980s, at the height of modernism, when nature and culture were radically separated, and the writing of history generally followed one or other of these themes, but not both.

Two centuries earlier, however, historians of the Earth were not so preoccupied with separating nature and culture, and were happy to map what Worster called "evolving relationships" between the two. The Comte de Buffon's scheme, outlined in his book *The Epochs of Nature*, influenced a whole generation of thinkers, including Alexander von Humboldt and Charles Darwin. Buffon sought to establish "the moment that the Earth became the domain of man", exploring human struggles with the land and, in particular, efforts to secure a "happy climate".[2] Buffon pointed out:

...man can modify the influences of climate in which he lives, and can secure, so to speak, the temperature to the level at which it suits him. And there is something that is singular: it is more difficult to him to cool the Earth than to heat it. Master of the element of fire, which he can augment and propagate at his will, he cannot do the same with the element of cold, which he can neither grasp nor communicate.[3]

Anthropocenic history has come full circle, returning to Buffon's world, his concern with heat and his efforts to achieve a global perspective.[4] In Buffon's time, "happy climate" was an ethnocentric concept: the best world was the one in which the observer lived. Now, the climate is far from "happy" and the planet struggles with excessive heat.

Several reports from agencies in both the United States and Europe conclude that the last decade was the hottest on record.[5] Records for 2019 indicated that average global surface temperatures were nearly 1 degree Celsius higher than the average from the middle of the last century. This state of affairs has largely been driven by human activity and the accelerating build-up of greenhouse gases in the atmosphere

Opposite above: Changes in global average temperatures.
Opposite below: Temperatures steadily rising from the late twentieth century onwards.

Temperature change in the last 50 years

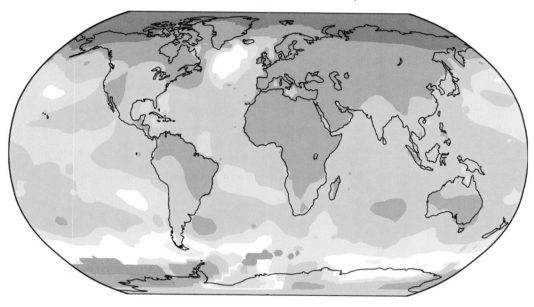

2010–2019 average vs 1951–1978 (°C)

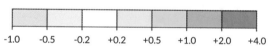

-1.0   -0.5   -0.2   +0.2   +0.5   +1.0   +2.0   +4.0

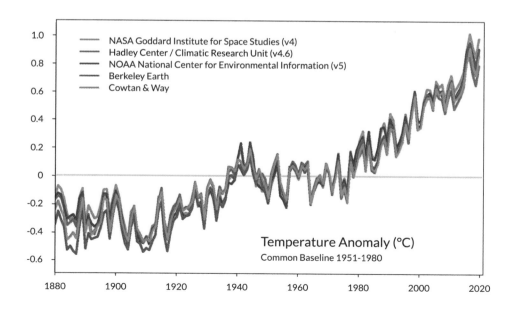

NASA Goddard Institute for Space Studies (v4)
Hadley Center / Climatic Research Unit (v4.6)
NOAA National Center for Environmental Information (v5)
Berkeley Earth
Cowtan & Way

Temperature Anomaly (°C)
Common Baseline 1951-1980

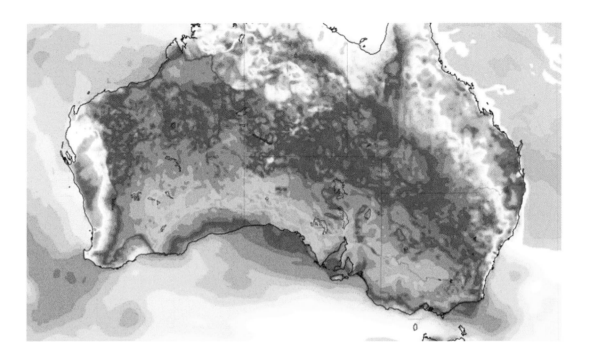

as a result of the burning of fossil fuels. Ocean records reveal similar prognoses, with significant likely impacts given that the ocean has absorbed at least 90 per cent of the extra heat generated as a result of anthropogenic climate change. A single degree may sound trivial, but it shows that the world is a long way from managing global heating and could be entering a dangerous "hothouse" state. Every step taken in this direction makes it more difficult for future generations to adapt.

Looking further back than the last decade, recent studies have confirmed the accuracy of global models covering the past 50 years, including models developed by former NASA climatologist James Hansen, the man who famously testified to the US Senate about anthropogenic heating.[6] Such findings increase the confidence that future climate projections will be reliable. This leads, in turn, to alarming conclusions. Judging from the

"Medieval Warm Period" between 1000 and 1200 CE, one of the catastrophic consequences will be extreme drought.[7] The next few United Nations Climate Change Conferences, including Glasgow 2020, will be decisive for global efforts to improve emission reductions, in line with the goals set out in the Paris Agreement of 2015.

"Overheating" is partly a metaphor, highlighting exponential speed and growth, multiplying energy needs, mass extinction, shocking pollution, etc. Norwegian anthropologist Thomas Hylland Eriksen points out in his book *Overheating: An Anthropology of Accelerated Change* that overheating, in the literal sense of excessive elevation of temperature, is part of a larger anthropogenic complex, involving globalization, tourism, pollution, shipping and neoliberal politics, which is difficult to disentangle. This intricate knot of factors makes the challenge of dealing with overheating difficult, chaotic and risky, with the potential to trigger mutually reinforcing "runaway processes", with unintended consequences that are hard to predict, understand and respond to.[8] Perhaps in the Anthropocene "superheating" would be a more apposite word than overheating; in physics, to superheat a liquid is to raise its temperature beyond its boiling point without allowing it to change phase, which can lead to it violently flashing to a gas.

Two centuries ago, the Comte de Buffon saw cooling as the project of civilizations – during the "seventh and last epoch" of Earth history "when the power of man has assisted that of nature". He would be baffled to learn of twentieth century air-conditioning machines. Humans have used billions of such machines for cooling dwellings and workspaces. Demand for them increases with global heating, which in turn exacerbates heating by increasing the demand for electricity derived from fossil fuels.

Opposite: Alarm call, heat record just a day after another record hit, Australia, 19 December, 2019.

Below: Portable fans on a hot day in Tokyo, August 2019.

Now, interestingly, there are attempts to use air conditioning-style machines to suck CO_2 out of the air during the process of cooling, producing climate-friendly, renewable "crowd oil" at the same time.[9] This is eco-science bordering on science fiction. In fact, eco fiction plays a growing role in human response to the climate crisis. All kinds of art, music, poetry and novels are helping people to move beyond graphs and statistics, to engage with the world as it is, and with possible ways of improving it.

Above: Air-conditioning ventilation units. The world now relies on these to keep them cool, yet they have an enormous carbon footprint.

THIRTEEN

The end of glaciers

Early glacial explorers usually thought of the massive ice sheets they encountered as inert and dead. This was a misconception, they later realized. Most modern glaciers were born and raised in the so-called Little Ice Age, the period of global cooling between roughly 1550 and 1850, and, in fact, they have always displayed seasonal variability. Glaciers that have retreated since the end of the Little Ice Age have left parallel scratches on rock faces, offering humans a lesson about the deep history of the planet, similar to that supplied by fossils. French historian Emmanuel Le Roy Ladurie found records in European archives describing a "horrible glacier of great and incalculable volume which can promise nothing but the destruction of houses and lands which still remain".[1] With the advent of global heating, the world's glaciers have now begun to teach humanity a second and more profound lesson, about the susceptibility of the Earth to human impact.[2] In some cases, melting is accelerated by particles from distant fires, caused by humans, which get trapped in snow and ice, darkening the surface of glaciers and increasing absorption of solar energy. In this way, recent fires and deforestation in the Amazon are contributing to the melting of glaciers in the Andes.[3] The main driver of melting, however, is

warming due to greenhouse gases, one of the key signatures of the Anthropocene.

The effects of glacier melting are manifold, and they will continue – albeit at different rates in different places – until the glaciers are gone. This will happen unless the global community manages to keep temperatures this century to no more than 1.5 to 2 degrees Celsius above pre-industrial (or pre-Anthropocenic) levels, in line with the goals of the Paris Agreement of 2015. The passing of glaciers invites a series of questions. What roles have the glaciers played in the past, in the lives of those who live or have lived in their neighbourhood? What are the local effects of melting? No doubt it is, to some extent, questions like these that explain why so many people today are choosing to visit glaciers, whether they are locals or come from farther away. Glaciers, in many places, are a treat for tourists and scientists.

One of the effects of glacier retreat is that the Earth has now begun a new dance around its axis. Melting of ice sheets is redistributing mass on the surface of the planet.[4] What the consequence of this will be is not quite clear and, in any case, it

Opposite: St. Mark's Square in Venice, October 2019, during the greatest floods since 1966.

THE END OF GLACIERS

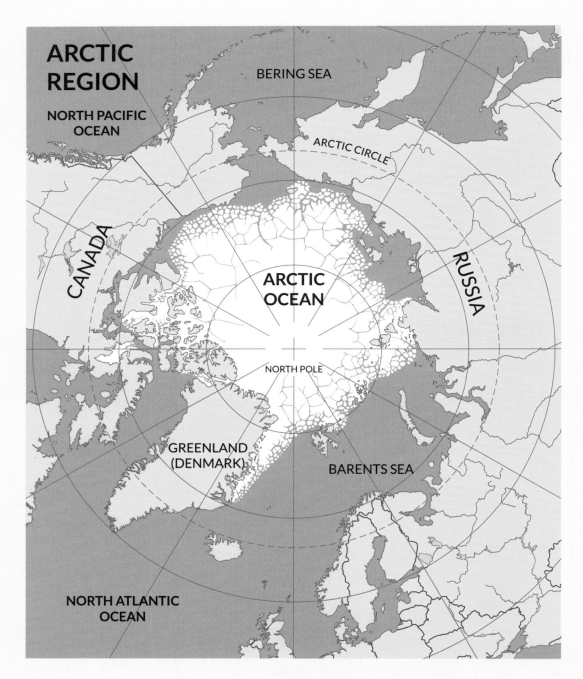

ARCTIC REGION

NORTH PACIFIC OCEAN

BERING SEA

ARCTIC CIRCLE

CANADA

ARCTIC OCEAN

RUSSIA

NORTH POLE

GREENLAND (DENMARK)

BARENTS SEA

NORTH ATLANTIC OCEAN

Above: The Arctic sea ice of Greenland is disappearing faster than predicted.

Opposite: *Lines (57° 59' N, 7° 16'W)*, 2018. Art installation exploring the rise in sea level, by Pekka Niittyvirta and Timo Aho.

The financial and psychological costs involved in adapting to floods (including "environmental grief") are enormous. The advancing threat is effectively captured in some visual artworks. One example is an installation in a town in Scotland, on one of the Hebridean islands, by Finnish artists Pekka Niittyvirta and Timo Aho. They have used sensors that interact with tidal changes, so that high tides activate synchronized light displays. The installation helps visitors to imagine future sea-level rise, provoking "a dialogue on how the rising sea levels will affect coastal areas, their inhabitants and land usage in the future".[11]

is much less noticeable than the more immediate consequences of flooding and sea-level rise, which are already forcing people to move, drowning historic settlements and threatening entire coastal cities with millions of inhabitants. In November 2019, the historic city of Venice, Italy, experienced its worst flood since 1966, as most of the lagoon city was engulfed. In the midst of the crisis, the Regional Councils of Veneto debated how to respond, eventually rejecting a plan to battle climate change. Minutes after the council made its decision, its offices were flooded.[5]

The melting of Greenland's ice is occurring faster than previously thought, with serious implications for Greenlanders and the planet. This is the world's second largest ice cap, after Antarctica, covering 80 per cent of Greenland. Its melting will contribute significantly to rising sea levels. For Greenlanders, the ice is vital, offering travel routes, hunting grounds and the keys to a traditional way of life, in the shadow of a market economy and colonial dependence (as part of the Danish state). Greenlanders consider their sled dogs to be family members; now they are having to kill them. The melting of sea ice triggers extreme anxiety among Greenlanders. In an interview with the *Guardian*, one hunter, Claus Rassmussen, expressed "a shocking explosion of grief", a highly significant response "in a culture where emotion is rarely shown" and where "strength, silence and self-sufficiency are admired traits".[6]

Glaciers, then, live and die. People who have lived in the neighbourhood of glaciers respect them and mourn their deaths. The Snow Star Festival (*Quyllurit'i*) in the Andes of southern Peru, a tradition that predated colonization, used partly to focus on celebrating the local glacier. During the annual festival, which took three days, men would dress as mythical half-man, half-bear creatures, cutting blocks of ice from the glacier to share with the community, believing that the melted water had healing powers for humans. Now, as a result of a decline in the size of the glacier in times of warming, this tradition has ended.[7] The loss of a local glacier takes its toll for the people involved. The mountains seem to weep.

The first "funeral" of a glacier took place in Iceland on 18 August 2019, when about 100 people from different parts of the world walked to the top of the mountain Ok, in West Iceland, a strenuous five-hour trek across extremely rough terrain, paying tribute to the first Icelandic glacier to melt down as a result, directly or indirectly (through blowing sand and particles), of global heating. A memorial plaque was placed on a mountain stone with the inscription "Letter to the future" written by Icelandic writer and activist Andri Snær Magnason: "We know what is happening and what needs to be done. Only you know if we did it". A few international dignitaries, notably Mary Robinson, former President of Ireland and United Nations High Commissioner for Human Rights, attended the event. It is easy to see the "ok-wardness" of the event – to coin an Anthropocenic term – to highlight the strange new normal of glaciers in times of melting, and their impact on the emotional attachments of humans to their planet.

Below: Quyllurit'i, the Snow Star Festival in the Andes of southern Peru. Men descend the Qullqip'unqu mountain glacier on 24 May, 2016.

Opposite above: The funeral congregation at Ok, Iceland, August 2019.

Opposite below left: Declaration of extinction, August 2019; the remains of glacier Ok in the background.

Opposite below right: "Letter to the future". Memorial plaque on Ok, Iceland, August 2019.

THE END OF GLACIERS

Ok Glacier is definitely dead – in geological reckoning, the remaining snow and sleet is no longer moving in a glacial fashion. This is not only the first glacier in Iceland to go and to be commemorated; its fate also immediately became a general symbol for glacier extinction, highlighted in hundreds of media headlines around the world. A month later a similar ok-ward moment occurred in Switzerland, with the mourning of the Pizol glacier in the Glarus Alps, Eastern Switzerland.[8]

Ash layers from volcanic eruptions are sometimes visible in the margin of melting glaciers, irregular horizontal stripes that time has engraved in the ice. Such layers provide important avenues into history.[9] It can look like the disappearing ice floes are poking out their "glacier tongues", in geological terms, to taunt observers. If things continue as predicted, the evidence of human actions will become ever clearer with each passing year, until the glaciers are gone.

Drilling down from the surface of a glacier, extracting "ice chronicles" from its depths, can offer highly valuable records of environmental change over thousands of years. On 1 July 1993, the Greenland Ice Sheet Project 2 sent a message to the world: "LOCATED IN CENTRAL GREENLAND … STRUCK ROCK. THIS COMPLETES THE LONGEST ENVIRONMENTAL RECORD … EVER OBTAINED FROM AN ICE CORE … AND THE LONGEST SUCH RECORD POSSIBLE FROM THE NORTHERN HEMISPHERE".[10] It was a historic moment. The project had established spectacular evidence about the distant past, samples of the atmosphere representing different points in time, and fundamental tools for future climatic projections.

Right: The mourning of a "dead" glacier in Switzerland, September 2019.

THE END OF GLACIERS

Results

Freakish weather

Climate and weather matter to most people most of the time, affecting mood and well-being.[1] In everyday life, the two terms are often used interchangeably, and both invite Anthropocenic interpretations. While they overlap, there is an important distinction. "Climate" – from the Greek *klima* (from *klinein* "to slope") – is usually seen as the long-term "behaviour" of the atmosphere; thus, global heating involves a temporal trend, a slope of some kind. "Weather", in contrast, refers to conditions over a short period of time. It can be calm and stable, or it can be freakish, extreme and violent, as in the case of recent firestorms, with dry heat and strong winds, or with devastating floods and hurricanes; all of these are regular events in some parts of the world. Climate, then, is "inclined" to change (both words have the same etymological roots), leading on to something, while weather is an event that "hits" – sometimes for the worse.

Below: Hurricane season: a warning sign.

Opposite: Houston, Texas, after Hurricane Harvey, 2017.

Previous: Mother and child in drought-ravaged Wajir, Kenya, 2006.

FREAKISH WEATHER

If we are able to influence climate, can we also make the weather, as the title of a recent book, *We Are the Weather*, would suggest?[2] How Anthropocenic is freakish weather? Is there a trend or a cline, too, in the case of extreme weather events? This question is difficult to answer. Ice cores and tree rings speak loudly on climate, but are less forthcoming concerning floods, and hurricanes. Also, the recording of temperature has a much longer history than the tracking of weather. The ability to engage at close quarters with hurricanes, for instance by flying into them, is a modern luxury for meteorologists.

Naturally, locals who have to live with freakish weather events have developed a complex vocabulary – including terms such as hurricane, cyclone, typhoon, tornado, flood, avalanche and tsunami. Such language is often documented in historical and ethnographic accounts from different parts of the world, based on the lived experiences of those who witnessed the events, but it is not preserved in the archaeological record of the deep past.[3] There is hope, however, for our historical quest and future prognosis. The 2007 Intergovernmental Panel on Climate Change concluded that tropical cyclone intensity had increased, even more than climatic models predicted, and that this trend was likely to continue throughout this century, adding that it is "more likely than not that there has been some human contribution to the increases in tropical cyclone intensity".[4] Other studies have concluded that while cyclone frequency may not increase, cyclones will become more extreme. In fact, the biggest cyclones are getting stronger. One study of Corpus Christi Bay in the Gulf of

Left: Houston, Texas, after Hurricane Harvey, 2017.

FREAKISH WEATHER

Mexico indicates that "if future global warming projections are realised, coastal flood levels have the potential to rise significantly over the next 80 years, making coastal communities progressively more vulnerable to hurricane damage".[5]

Hurricane intensification means elevated flood risk, posing serious threat to coastal communities and human lives. Recent historically destructive hurricanes include Katrina (2005), which devastated New Orleans and surrounding areas and claimed 1,200 lives, and superstorm Sandy (2012), which killed at least 230 people in eight countries, with vast emotional, political and financial implications.[6]

The 2019 Atlantic hurricane season was close to or above the average for the last decade, in all hurricane metrics. This was "the fourth consecutive year in which a category 5 storm

developed in the Atlantic basin – a new record", as Jeff Masters points out in his chronicle of the hurricanes of the 2019 season.[7] "Are they", he wonders, "harbingers of the future?" One of them was the ultra-intense Dorian, with 185 mph winds that devastated the Bahamas, the southeastern United States, the Virgin Islands and even parts of Canada, causing $4.6 billion in damage.

Attempts have been made to weaken and slow down hurricanes. In the US Government's Project Stormfury, in the United States in the 1960s, 1970s and 1980s aircraft were flown into selected storms to seed them with silver iodide, in the hope that this would supercool water in them

Opposite: In the eye of the storm of Hurricane Dorian, 1 September, 2019.

Above: Houston, Texas, after Hurricane Harvey, 2017.

to the point of freezing, slowing them down and perhaps diverting their paths. It was discovered, however, that there was too much uncertainty with the outcomes. Another ambitious proposal involved cooling the water below a cyclone by towing icebergs into the tropical oceans. But tropical cyclones seem to be too powerful for humans to "manage". We may be able to intensify cyclones, but once the storms get going they are beyond the reach of humankind. It makes more sense to develop disaster-mitigation measures, setting up defences and addressing problems as they arise during and after storms.

It is interesting to note that in some contexts where damaging hurricanes are routine, governmental and state relief seems conditional on not mentioning climate change.[8] In the United States, billions of dollars are allocated to coastal states for preparations in advance of hurricanes and for mitigating their impact, apparently on the condition that the elephant in the room is not specifically addressed. Thus, a 306-page proposal from the state of Texas neither mentions "climate change" nor "global warming", citing only "changing coastal conditions". In other cases, official reports and rules governing the use of disaster funding acknowledge "changing weather conditions" and rising seas, without mentioning climate change.

The current American discourse around climate disasters as a secret thing, almost a war on facts, is partly a result of the Cold War. The atomic arms race led to a state obsessed with surveillance and national security. It had become possible, for the first time, "to imagine a truly planetary crisis".[9] While cyclones are a fact of life in some contexts, discussions of the potential causes of their intensification as a result of human activities

Above: The crew of Project Stormfury and one of their Douglas DC-7 aircraft, 1966.

Below: Still from the film *The Day After Tomorrow* (2004).

have been carefully contained and domesticated. Climatologist James Hansen was subjected to hacking and harassment after his warnings about the future threat of global warming, and access was severely restricted to scientific reports exploring the connections between climate change and the escalation of hurricanes.[10] The technology that is used to predict and track cyclones draws heavily upon advances generated by the Cold War, including detailed mapping of the Earth, supercomputers and satellite surveillance systems. It was not a coincidence that public officials likened Hurricane Katrina to nuclear war and the bombing of Hiroshima. Nature offered new "wars on terror".

Cyclones and nuclear winters have caught the public attention through various arts and media, including films, books and music. Some of the popular films are *The Matrix*, *The Perfect Storm* and *The Day After Tomorrow*. In the latter, Los Angeles is destroyed by tornadoes and New York is flooded by rising seas and later frozen solid, while the security state refuses to listen to climate scientists and their warnings about global warming. Together, these films probably give a sense of apocalypse and despair, but perhaps they can open a space for readiness and hope, and a constructive collective response, bracing us for the floods and cyclones ahead.

Volcanic eruptions

Gradually, scientists added earthquakes, volcanoes and geological strata to the scientific map of the world. "What causes eruptions?" they asked. To some, the answer seemed obvious once the theory of continental drift was advanced by German meteorologist Alfred Wegener in 1912. Continental drift, now called plate tectonics, postulates that the Earth's crust consists of a number of thick plates that are in slow but constant motion above the furnace at the centre of the Earth; they move apart or together, or slide under one another. Where the earth opens

between plates, it gives rise to ridges, volcanoes and eruptions. Around the mid-1960s, the theory of plate tectonics became a widely accepted truth. However, it was not until the 1990s that people monitored plate movement directly. With the help of satellites, it became possible to measure Earth-based movements from space.

How deep is the human impact on the planet? Could Anthropocenic signatures be detected in the interior, the subterranean world of plates and magma? The answer is yes. Fracking, a process in which liquid is forced into the earth

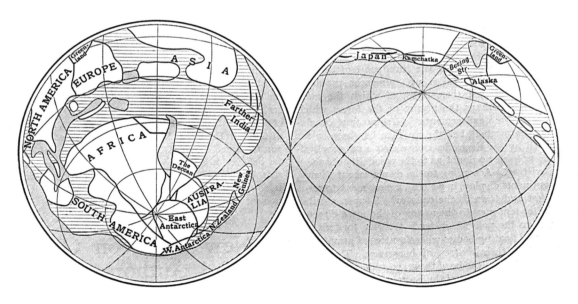

to facilitate removal of oil or gas, has initiated earthquakes that would not otherwise have occurred.[1] Utilization of geothermal energy has a similar effect. What about volcanoes? Are they immune to human influence, forces of nature in the classic sense, remnants of the untouched and pristine – along with extraterrestrial phenomena, including the Northern Lights?

Eruptions can have dramatic effects. The eruption of Eyjafjallajökull in South Iceland in 2010 is a case in point. Thanks to satellite technology, the progress of the eruption could be watched live around the world. Soon, Iceland was on everyone's lips. Eyjafjallajökull (jökull meaning "glacier") closed airports across Europe.

Since volcanic ash can cause engine failure, planes could only fly when the wind direction was favourable. The closures had a considerable impact on daily life and travel for 10 million air passengers and their families all over the world, far from the actual volcano.[2] For a brief period, it slowed down the emission of greenhouse gases by grounding airlines, producing lots of its own gases in the meantime.

The Laki eruption in Iceland, 1783, occurred at the dawn of the Anthropocene and there is no

Opposite: Diagram of the Earth during the Carboniferous period, from an article by Alfred Wegener on his 1915 theory of Continental Drift.

Above: The eruption at Eyjafjallajökull, Southern Iceland, 2010.

indication, so far, that it was the result of climate change. However, a historical study of eruptions in Iceland in the middle of the Holocene, about 5,500 to 4,500 years ago, shows that glacial expansion puts pressure on the Earth's surface, slowing down eruptions. Could a recent upsurge in eruptions be considered human-induced, a result of warming and glacial retreat?

Indeed, some Arctic volcanic eruptions have been attributed to human activities.[3] A study from 2008 of Iceland's Vatnajökull ice cap indicates that melting glaciers can increase volcanic activity over time frames that are relevant to humans.[4] Another recent study of Icelandic volcanoes concludes that uplifting and volcanic activity are probably the result of global heating.[5] Drawing upon continuous measurements at 62 points in Iceland, using Global Positioning System (GPS) technology, the research team (based in Arizona) concluded that faster uplift of the Icelandic crust (as much as 35 mm per year) has coincided with eruptions and the onset of heating since 30 years ago. The *Guardian* appropriately claimed: "Climate change is lifting Iceland – and it could mean more volcanic eruptions". Some media described the Icelandic crust as like a trampoline, in constant motion. It seems probable that the Eyjafjallajökull eruption of 2010 was triggered by heating, ice loss and decompression. This is an important new twist on the Anthropocene. Welcome to the trampoline!

Although Earth scientists are still unable to travel far beneath the surface of the Earth, they can now look deep into our dynamic planet. With the assistance of supercomputers

Left: Hand-coloured engraving by one of the first volcanologists, Sir William Hamilton, of the crater of Mount Vesuvius before 1776.

and mathematics, they can "scan" the interior of the Earth as though it were a human body, and monitor long-distance magma movements deep beneath the surface. The comparison with the human body is not trivial. As archaeologist Karen Holmberg argues, people often anthropomorphize volcanoes, attributing fingers, shoulders, necks and even personalities to them; few volcanoes carry more symbolic weight than Mount Vesuvius, thanks to its simultaneous destruction and preservation of the first-century Roman town of Pompeii.[6] Mount Vesuvius has often been represented as a living agent, randomly opening up and shrugging its shoulders. Such an understanding is vital for hammering home the point that the material volcano is very much alive, highlighting at the same time that people often feel related to "their" volcano; understanding its acting "body" is important for those who live with it. A volcanic eruption above or nearby human settlement can be seen as a useful metaphor for the disruption of an established order, illuminating the key problems of life and the necessity of concerted action on a larger global scale.

An installation, *The Other Volcano* by French artist Nelly Ben Hayoun, makes this point wonderfully, illustrating the dynamics of human-induced eruptions during the Anthropocene right at home, in living rooms. In 2010, the artist placed a few man-made volcanoes in Londoners' living rooms. She had modelled her volcanoes on "live" examples: Mount St. Helens in the US and Ol Doinyo Lengai in Tanzania (the name comes from the Masai for "Mountain of the Gods"). She inserted detonators and explosives from fireworks in the interiors of her volcano sculptures and had them erupt now and again. Volunteers "took care" of them at home for two weeks and became experimental subjects in the process. Ben Hayoun describes her installations on her website:

> The Other Volcano *imagines a love-hate relationship, a "sleeping giant" in the corner of your domestic environment, with the power to provoke excitement with its rumblings, and also perhaps fear (if not for one's life in this case, then at least for the soft furnishings of one's clean and neat 'living' room).*[7]

Geographie der Pflanzen in den Tropen-Ländern;

Ben Hayoun's experiment is about threatening the cosy space of the living room with the unpredictable performances of a heaving, living mountain – rather like having an exotic guest in the house. Global heating is a giant volcano in the room. And so are we, considering the evidence for human-induced eruptions.

Human impact on the interior of the planet, through fracking, drilling and the manufacture of earthquakes and eruptions, may seem relatively trivial, compared to that of greenhouse gases through travel and industry. It is difficult, however, to disentangle "single" Anthropocenic effects. In fact, one can expect a combined spiralling effect of various Anthropocenic factors: emissions of greenhouse gases lead to extreme heat, which melts glaciers, leading to uplift of the planetary crust, which generates earthquakes and eruptions, which result in more greenhouse gases – and so on. This spiralling can only accelerate with recent thermal extremes.

After all, everything planetary is interconnected – as Alexander von Humboldt hinted at early on, with his stunning works on volcanoes and warming. Von Humboldt's representation of the interconnectedness of nature in his *Naturgemälde*, or "painting of nature" (also called the Chimborazo Map after the volcano Chimborazo), often used for analyses of global heating, was a pioneering infographic advance, introducing isothermal lines and packing in different kinds of environmental factors, including volcanoes.

Opposite: Elliðaey, one of the Westman Islands south of Iceland. The flat oval-shaped rock is the top of a crater which erupted thousands of years ago, producing the tiny island surrounding it. The cruise ship in the distance is a reminder of the geologic impact of international travel and global heating, possibly generating earthquakes and eruptions, opening the lids of "extinct" volcanoes.

Above: Alexander von Humboldt's *Naturgemälde*, or "painting of nature", also called the Chimborazo Map, describes a volcano (Chimborazo) and plant geography.

Following pages: Eyjafjallajökull, Southern Iceland, 2010.

The fragile ocean

The ocean is vital to the history and future of life. It was the birthplace of the earliest organisms and it remains an essential storehouse of water for the biosphere of the entire planet (hosting 97 per cent of global water), regulating carbon and climate cycles. For too long, however, humans have viewed the ocean from its edges, looking out from the liminal shore.[1] During the colonial era, the human view of the ocean was vastly expanded, opened up by European fleets and explorers, slave traders and plantation owners.[2] This marked the beginning of the Long Anthropocene.

The shoreline and shallow waters have always been a source of food and salt. Historically, salt was seen as a positive thing, a means of conserving food and preserving the physical remains of the dead.[3] Like money, salt seemed powerful, imperishable and quantifiable – an observation preserved in the etymology of "salary". In ancient Rome, labourers extracted salt from the sea and transported it to the city, in return for a *salarium*: "salt-money", or salary. In Biblical language, some people were more "worthy of salt" than others. Salt was not always a blessing, though. Apparently, in the Middle Ages salt was spread on land to poison it, to punish landowners who had violated the interests of the collective.

In the Western imagination, the ocean itself remained two-dimensional, an immense seascape for travel, commerce and communication. The fascination with ocean monsters – including the Leviathan of the Hebrew Bible, the Midgard Serpent of Norse mythology, and the giant white sperm whale of Herman Melville's novel *Moby Dick; or, the Whale* – was more a metaphorical reflection of the social world of humans than a keen study of life in the ocean depths. Humans had minimal interest in the watery underworld, a substantial part of the planet, mainly because of the difficulty of access. The heavens and the planets were more accessible for observation and far more interesting.

Below: *The Sea Serpent*, engraving from *Journal des Voyages*, 1879–80

Nineteenth-century Westerners developed an increasing interest in the ocean, often through the construction of aquaria, which became public sensations, especially in Victorian Britain, offering the chance to gaze into a mysterious world normally hidden from view.[4] It is pertinent to speak of the *birth* of the aquarium as a general model of the oceans. An illustration by artist Bruce McCall, *Lobsterman's Special* – lobsters ordering their dinner from a tank of swimming humans – helps to illustrate some of the implications of the aquarium.[5] Above all, the oceans tend to be seen as a gigantic fish tank, scientifically managed for human purposes. One species – humans – enjoys the position of the observer and manipulator while other species (fish and other aquatic animals) occupy a secondary position. Implicit in that distinction is the topological separation of the inside of the aquarium and the outside world, and the related distinction between practitioners and experts. Images of aquaria have often been used to convey a sense of the necessity of "building" a rational world for future generations. By reversing the roles of humans and lobsters, the artist seems to call attention to our implicit division of the human and the natural into separate realms.

Until fairly recently, Westerners typically assumed that the supply of living resources in the ocean was a boundless one. Such a position, of course, was untenable in the long run. Indeed, the ocean and its resources, especially fishing stocks, played an important role in early theorizing about the commons, resource management, and carrying capacity.[6] Many of the world's major fishing stocks are threatened by overfishing, global heating and pollution – from oil, radioactive waste and other by-products of human activities – and fisheries more and more resemble other branches of industry. For one thing, the boundaries of "wild" fisheries are increasingly blurred, with exponential growth in sea ranching and fish farming.

In the 1950s, Rachel Carson, better known for her work on the use of pesticides in agriculture, drew attention to the ways in which ocean life was affected by warming waters. She wrote in her book *The Edge of the Sea*: "the level of the sea itself is never at rest ... [it] rises or falls as the glaciers melt or grow, as the floor of the deep ocean basins shifts under its increasing load of sediments, or as the earth's crust along the continental margins warps up or down in adjustment to strain and tension".[7]

The ocean is severely affected by global heating and increasing emissions, sucking up about 25 per cent of all the extra CO_2 emitted. As it warms it absorbs even greater amounts of carbon, becoming more acidic, threatening ecological collapse. "Dead zones" in the ocean, where oxygen is effectively absent, affect many species, especially large fish, including tuna. This has serious consequences for terrestrial life as well, which to a large extent depends on oxygen generated by the ocean. These processes are slowly being understood through detailed longitudinal studies covering the last three decades, across different parts of the world.[8]

The results of such studies, along with evidence of prehistoric volcanic eruptions and meteor impacts, help to predict future scenarios around CO_2 emissions and warming, precipitation and weather, and the chemical composition of the ocean. "Ocean acidification" is occurring faster than ever before in history, threatening mass

Opposite: The appearance of sea monsters in *Liber Floridus*, a medieval encyclopaedia compiled between the years of 1090 and 1120.

extinction of species which have co-evolved over a very long time under radically different conditions. Reduced calcification rates threaten reef-building and corals, with unprecedented environmental changes in coastal areas, among the most threatened ecosystems of the planet.

The Atlantic Ocean harbours a giant "conveyor belt" of currents, which moves water both horizontally and vertically, with important implications for marine life, ocean chemistry and global weather patterns. These currents are not immune to global heating. Again, there is a lot to learn from prehistoric events (calibrated with the help of volcanic ash and ice cores). It seems that combined studies of the deep past and recent trends offer a warning sign that the conveyor belt is slowing down. Although the impact of ocean warming is not always immediate (lag times

Below: Saint Joseph Atoll in the Seychelles, a nature reserve with a marine protected area.

Opposite above: Funafuti Atoll in the west-central Pacific Ocean is only 15 feet above sea level at its highest point.

Opposite below: The "conveyor belt" currents of the North Atlantic Ocean.

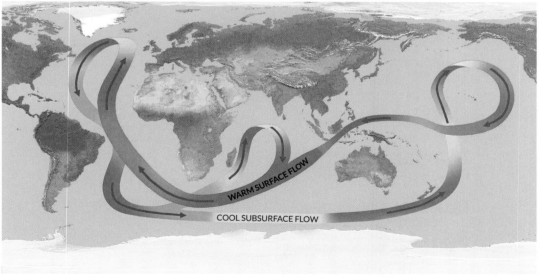

WARM SURFACE FLOW

COOL SUBSURFACE FLOW

HIGHEST AND DEEPEST POINTS ON EARTH

Mount Everest
(Chomolungma)
The Earth's highest mountain.

Mount Everest	Mount McKinlay	Mount Sharp	Mount Rainier
8.8km	6.2km	5.5km	4.4km

First ascent
29 May 1953
Edmund Hillary
and Tenzing Norgay

✺ **Location**

Mahalangur section of the Himalayas,
China and Nepal.

8,848 m
Mount Everest

5 km
4 km
3 km
2 km
1 km
0
1 km
2 km
3 km
4 km
5 km

10,911 m
Mariana Trench

✺ **Location**

Western Pacific Ocean, to the
east of the Mariana Islands

Mariana Trench
The deepest part of the world's oceans.

Descents

1960	1995	2009	2012
Trieste (USA)	Kaiko (Japan)	Nereus (USA)	Deepsea Challenger (USA)

Above: The scaly-foot snail, *Chrysomallon squamiferum.*

Right: *Trieste* (designed by Auguste Piccard in 1960), one of the vessels aimed for Mariana Trench (named after the Mariana Islands in the Western Pacific).

Opposite: The highest and deepest points on Earth: Mount Everest and the Mariana Trench.

being in the order of centuries), there could be exacerbation of hurricanes.[9]

While the ocean has been thoroughly explored for over a century, the sea bottom is still poorly understood. One reason is the difficulty of access, given the immense pressure in the depths, and until recently, human explorers saw no point in making the effort to get there. Ecosystems, it had been assumed, must be based on photosynthesis, which is impossible thousands of feet below the sea surface.[10] This view changed in 1977 with the discovery, on the sea floor near the Galápagos Islands, of hydrothermal vents teeming with life, including crabs, worms and octopuses that thrived in an ecosystem based on chemicals coming from the vents. The deepest place on Earth is the Mariana Trench, at a depth of about 11,000 metres (36,000 feet). It already has plastic bags.

Currently, a number of companies are scrambling to exploit rare and valuable metals from the deep sea bed, before international regulations are put in place. Underwater mining,

already planned or taking place within the coastal waters of Namibia, Papua New Guinea and several other places, can ruin delicate ecosystems, which are often minimally documented and understood. One of the strange creatures found around deep-sea hydrothermal vents is the scaly-foot snail, *Chrysomallon squamiferum*, discovered in the Indian Ocean in 1999. It has a strange multi-layered metallic armour made of iron sulphide.

The aquarium may, after all, remain an effective environmental metaphor, as long as one recognizes the interconnectedness of everything in the biosphere. For such a metaphorical association to make sense, it would be necessary to relax some assumptions about control and boundaries. An updated aquarium metaphor would have to include climate scientists as well as "laypeople", swimming, along with whales, fish, microbes and other earthlings, in the tank of life. Such a reassessment is particularly important in the Anthropocene, with its human signatures at every level of the living and material world.

SEVENTEEN

....

Social inequality

Does the Anthropocene affect people equally, independent of wealth, social position, race, gender and class? The question has been a matter of some debate, often with reference to the "lifeboat argument" of Dipesh Chakrabarty, a highly influential Indian historian and social theorist based at the University of Chicago. Chakrabarty argued in 2009:

Climate change, refracted through global capital, will no doubt accentuate the logic of inequality that runs through the rule of capital; some people will no doubt gain temporarily at the expense of others. But the whole crisis cannot be reduced to a story of capitalism. Unlike in the crises of capitalism, there are no lifeboats here for the rich and the privileged ...[1]

Chakrabarty emphasized, while acknowledging inequality, the simple fact that no one can leave the damaged Earth. In that sense the Earth is treating "us" as equals; since we can never

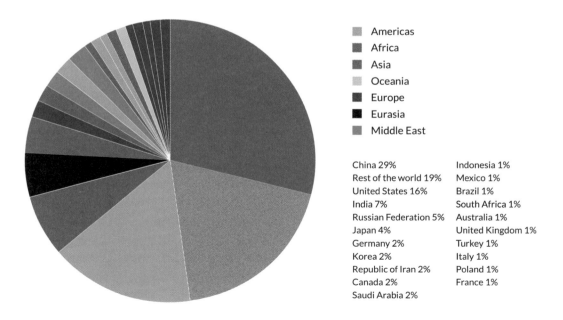

Americas
Africa
Asia
Oceania
Europe
Eurasia
Middle East

China 29%
Rest of the world 19%
United States 16%
India 7%
Russian Federation 5%
Japan 4%
Germany 2%
Korea 2%
Republic of Iran 2%
Canada 2%
Saudi Arabia 2%

Indonesia 1%
Mexico 1%
Brazil 1%
South Africa 1%
Australia 1%
United Kingdom 1%
Turkey 1%
Italy 1%
Poland 1%
France 1%

leave, lifeboats are trivial. Among the critics of the lifeboat argument were Swedish human ecologists Andreas Malm and Alf Hornborg. For them, this was flawed reasoning, overlooking the differential impact of recent disasters (including that of Hurricane Katrina) on human society, with some groups of people being extremely vulnerable while others, the privileged, were protected: "For the foreseeable future – indeed, as long as there are human societies on Earth – there will be lifeboats for the rich and privileged. If climate change represents a form of apocalypse, it is not universal, but uneven and combined".[2] In a later piece, Chakrabarty emphasized humanity's devastating ecological footprint, which is leading to denial of citizenship and statehood to the homeless, environmental migrants and refugees.[3]

While Malm and Hornborg, perhaps for the sake of argument, exaggerated the difference between their position and that of Chakrabarty, they made a couple of broader observations about the "social". For one thing, Anthropocenic effects on social life are manifold, and they should not be left to the geoscientists who typically lead major discussions of global heating, as they are not trained to be sensitive to the nuances of societal issues. Also, Malm and Hornborg observed, while the fate of humanity and the planet are increasingly intertwined – in a "geosocial" fashion, as some would have it – it may be necessary for observers to maintain some distinction between the social and the geological. Indeed, a number of social scientists

Opposite: Each country's share of CO_2 emissions.
Above: Bartolome in the Galápagos Islands, an elite resort.

SOCIAL INEQUALITY

and humanities scholars have been addressing some of the key issues in recent decades.[4]

Currently (as of 2019), China and the United States are the biggest emitters of greenhouse gases, accounting for almost half of global emissions. Not only do some countries bear greater responsibility than others, but some individuals account for a disproportionate carbon footprint. In January 2020, the *Guardian* reported on "'no emission spared' round-the-world holidays in private jets".[5] Apparently, 50 members of the wealthy elite were planning to board a privately chartered Boeing 757 for 10 flights and 23 nights "in five-star hotels or lodges, meals in some of the world's most famous restaurants, champagne". These are luxury lifeboats, but they will not operate for long: they have fewer places to visit, and they will only get away from it all momentarily. Differences in wealth are an immense Anthropocenic problem. Excessive wealth tends to translate into massive environmental impact, even ecocide (the destruction of the natural world), blinding people to the realities they face. One of the by-products of excessive wealth is the funding of "climate scepticism" and opposition to carbon taxes. Arguably, the world needs a ceiling on personal income, as well as a poverty line, and strict legal frameworks that make ecocide a punishable crime.

Interestingly, the term "wealth" originally meant (in Middle English) "well-being" or "wellness", closely aligned with happiness and health.[6] Given the environmental footprint of excessive riches, the meaning of "wealth" has now been

Left: Pemba, Mozambique, 2019. Two young men walk in the aftermath of floods caused by Cyclone Kenneth.

sharply derailed, and now seems to encompass "environmental injustice", misery and destitution for others. In 2019, the United Nations warned of a "climate apartheid" as the Earth continued to break new heat records, with a divide between the most vulnerable and those who can afford to escape, a division that is driving millions of people into poverty, with immense implications for human rights, living conditions and democracy.[7] Many people are stuck in devastating floods, while others die of excessive heat.

Feminist scholars have challenged both the masculinist bias of Anthropocenic discussion (the "Manthropocene", as some have called it) and the disproportionate effects of climate change on women.[8] Among the poor, women are often the most vulnerable in times of oppression and disaster, prey to vicious cycles of poverty, malnutrition and pollution. The violence affecting many women is liable to be ignored, and medical attention is likely to focus almost exclusively on the pregnant belly and its contents.[9]

Social and cultural differences are registered in the lives of the present generation, inscribed in the human body, in much the same way that language and accent are imprinted through upbringing. Gender, race, and social class, it is often argued, are both social and biological phenomena, embodied and reproduced from one generation to another. If the body has "biosocial"

memory of this kind, independent of genetics, it is likely to memorize health, poverty, abuse and trauma. Epigenetics is a growing field that explores transgenerational inheritance at the molecular level of the chromosome.

One of the most cited studies involving the effects of foetal environment, concerns pregnant women affected by the Dutch Hunger Winter of 1944 to 1945, the effects of which were tracked for two ensuing generations. Thirty thousand people died from starvation in the Netherlands as a result of a food embargo imposed by the Germans in the Second World War. Birth records collected since that time have shown that children born of women who were pregnant during the famine not only had low birth weights, but also exhibited a disproportionally high range of disorders later in life, including diabetes, coronary heart disease, breast and other cancers.[10] Epigenetics is still a contested field, but if the evidence of the Hunger Winter and similar studies testifies to transgenerational inheritance, one can imagine that the impact of the unfolding Anthropocene on future generations will be colossal, particularly in areas and families worst hit by pollution and poverty.

Opposite: Women and children in disaster zone.

Above: Dutch children eating soup during the famine of 1944 to 1945.

Anthropocenes, North and South

The Anthropocene, as we have seen, has been incredibly productive in generating conversations about environmental change at a critical moment in human history. Many social scientists and humanists, however, have challenged unqualified uses of the term, and rightly so, pointing out that current environmental crises are not the result of the actions of an undifferentiated humanity, but of particular people at particular points in time.[1] Environmental literature (including this book) frequently refers to sentences of this kind: "We have fundamentally changed the planet" and "Our signatures are (or will be) visible everywhere". It is pertinent to ask: who are "we"?

It is important, indeed, to attend to radical differences between us, in particular with respect to social inequality, geography, race and gender. Current crises are the product of a particular segment of humanity and a specific economic formation – that of industrial capitalism. Some propose that the current epoch should be called the "Capitalocene" rather than the Anthropocene, but, again, this would suggest an undivided capitalism, without distinguishing small-scale production in impoverished rural communities throughout the world.[2] Rather than focus exclusively on alternative terminologies, some of which are unlikely to gain traction anywhere (for example, "Cthulucene"), it might be more productive to explore histories of involvement and responsibility, as well as divisions within the human community at different scales.

As British anthropologist Chris Hann has argued, it is important to complement current anthropological descriptions of everyday life "with a truly *longue durée* account of how our planet has come to be where it is today".[3] It makes sense to speak broadly of different Anthropocenes, with different trajectories informed by centuries of geographic, racial and cultural differentiation and tension. For one thing, industrialization in Europe meant enforced social division, driving peasants from the land to establish stable labour forces in factories and expanding towns. These people and their descendants are hardly responsible for the problems of the Anthropocene to the same

Opposite: A Somali refugee mother and child, Hagadera refugee camp, part of the giant refugee settlement in Dadaab, Kenya, 2011.

ANTHROPOCENES, NORTH AND SOUTH

ANTHROPOCENES, NORTH AND SOUTH

extent as nineteenth-century aristocrats and factory owners. More importantly, the driving force behind Anthropocenic industrialism, the extraction of fossil fuels and the growth of greenhouse gases, was that of European colonialists who subjugated the New World to their domination, capturing millions of people from Africa as slaves for plantations in the Americas. Former colonies in what used to be called the developing or the Third World (now sometimes called the Global South) are hardly responsible for the current crisis to the same extent as the so-called developed or First World of formerly colonial powers or the Second World, the nations "behind" the iron curtain of the Cold War.[4] A person in Africa who practices a mix of agriculture, hunting and gathering does not bear the same responsibility for climate change as an American CEO of an oil fracking company.

Given the colonial roots of the Anthropocene, the most obvious global geographic divide is that of a Northern and Southern Anthropocene: the latter would include Africa, Asia and Latin America, each with its own histories and characteristics. Emphasizing that the Anthropocene must feel different depending on location, American anthropologist Gabrielle Hecht explores what an African Anthropocene might look like: "What picture of the Anthropocene ... emerges when we begin our analytic adventure in Africa instead of Europe?"[5] African minerals helped to drive industrialization and the development of nuclear weapons, leaving clear signatures in the geological record for millennia. South African

Left: Waste in Lagos, Nigeria.
Overleaf: A lake of garbage in Lagos, Nigeria.

ANTHROPOCENES, NORTH AND SOUTH

workers lugged rocks containing uranium to the surface, many of them losing their lives in the process. "The term did not yet exist", Hecht points out, "but the Anthropocene was nevertheless etching itself into the lungs of generation after generation of young African men".

When South Africa's apartheid government launched its uranium-processing plant in 1952, it set in motion an environmental disaster not often flagged in Anthropocene literature.[6] Heavy metals, including arsenic, mercury and lead, formerly trapped in rock, were now freed. As they dissolved in water, they made a toxic soup which continues to affect thousands who use the water for drinking and bathing. This is how the Anthropocene looks for poor South Africans.

Anthropocenic bodily effects are not only selective with respect to gender and social class, they also have a clear North-South registry. One of the African cases discussed by Hecht is that of air pollution.[7] Diesel vehicles that are dumped in Europe due to tight emission restriction have a second life in African cities, among them Lagos and Accra. The fumes from "dirty diesel" hovering above Lagos contain 13 times more particulate matter than London, with obvious consequences for the health and life expectancy of the inhabitants. Hecht encourages people "to think from, and with" Africans: "'They' are 'us', and there is no planetary 'we' without them".

African political scientist Abdirashid Diriye Kalmoy has a similar take on the African Anthropocene. While Africa, he points out, is endowed with rich natural resources and cultural traditions, it is "the epicentre of the Anthropocene:

Below: Grootvlei, Snake Park: a suburb on the fringe of one of the biggest mine dumps in Soweto, Johannesburg.

Opposite: Diesel pollution in Accra, Ghana.

ANTHROPOCENES, NORTH AND SOUTH

ANTHROPOCENES, NORTH AND SOUTH

Opposite: Saint-Louis, Senegal, is in a permanent state of flood alert.

Right: Langue de Barbarie, adjacent to Saint-Louis, Senegal.

decimation, destruction and plundering of its ... landscapes are normalised".[8] While modern-day colonization continues, he goes on, global environmental meetings typically express "indifference and apathy towards Africa's dooming Anthropocene". If this is the case, it probably reflects the lack of attention to race in geoscience institutions, which are typically associated with rugged white males. Addressing the issue, Kuheli Dutt, assistant director for academic affairs at a major geoscience institute (the Lamont-Doherty Earth Observatory at Columbia University, New York), concludes: "A lack of diversity and inclusion is the single largest cultural problem facing the geosciences today, and this is probably not just limited to the United States".[9] Given this state of affairs, it is perhaps not surprising that global environmental fora fail to address the African Anthropocene, and probably the Asian and Latin American ones as well.

One striking remainder of the Eurocentrism of current environmental discussions is that while the recent floods of the historic city of Venice, Italy, have been extensively covered in international media, the rising seas affecting the "Venice of Africa", Saint-Louis in Senegal, have received scant notice.[10] A historic colonial centre of Senegal, Saint-Louis has been on permanent alert and is now more or less gone. Many coastal cities of Western Africa, home to millions of people, face a similar threat.

As the Anthropocene is the product of specific historical economic forms, particularly late twentieth-century neoliberalism, mitigating Anthropocenic damage and reducing global environmental change will require changes to dominant economic models and financial practices – a new world order. Above all, this world order must allow for more equity in terms of race and geography, in addition to reducing the symptoms and repercussions of the Anthropocene.

The sixth mass extinction

A recent report by the United Nations, "One million threatened species?", suggests that discussions of the current environmental crisis should not focus only on global heating, and that decreasing biological diversity is a growing and serious threat.[1] In Earth's history, there have been five previous waves of mass extinction, as a result of meteor impact or volcanic eruptions. Now a new wave of extinction, apparently the sixth, this time driven by humans, is unfolding. A million species of plants and animals are on the brink of extinction, threatening to eliminate a "significant proportion of the world's biota in a geologically insignificant amount of time".[2] If a million species disappear – a significant part of the total number of known species – the entire ecosystem of the planet will face irreparable damage. In the 1980s, British geographer I.G. Simmons estimated that the "normal" or background extinction rate, as part of evolutionary change, was 1 per cent of the total number of existing species every 2,000 to 3,000 years.[3] The contrast between this and the rate (1 million of approximately 8.7 million known species) highlighted in the UN report is shocking. This is the scale of the Anthropocene extinction at human hands.

Western thinking, from Aristotle onward, tended to assume that nothing in nature was without justification, and nothing would disappear, least of all humans, unless it was obviously devoid of purpose. Also, it should be noted, Western thought took for granted that ethics and justice, the hallmarks of civilization, were ingrained in the natural world. As a result, extinction was unlikely, if not impossible. Species, it was understood, might drop out of view for a while, moving from one region to another, but they would not disappear for good. It was not until the eighteenth century, especially in the work of philosopher Immanuel Kant, that these ideas were challenged. Kant argued that morality and ideas about justice were human constructions, not parts of the natural world. If humans were not careful, they might disappear entirely. Other factors would also challenge received wisdom

Opposite: Endangered Hawaiian monk seal caught in fishing tackle off Kure Atoll, Pacific Ocean.

Overleaf: Sea ice is integral to the survival of polar bears.

THE SIXTH MASS EXTINCTION

THE SIXTH MASS EXTINCTION

about nature and the human condition: the world had expanded with the discovery of the New World and industrialization, developments which called for new ideas.

Homo sapiens is not exempt from the prospect of mass extinction. Quite possibly, *Homo sapiens* – once considered the master of the planet – will vanish from the Earth, a place it has helped to shape through its tools and intellect. Carl von Linné granted *Homo sapiens* a place in his grand classificatory scheme in 1758. French philosopher Denis Diderot suggested a little later that *Homo sapiens* would probably become extinct eventually but the species would, at some point, re-emerge.[4] In 1836, Italian writer Giacomo Leopardi maintained, around the time of the disappearance of the great auk, that if the human species were to leave, "the Earth would not miss a thing". The prospect of human extinction is increasingly on the agenda. Will anyone miss us? Does it matter if no one is around to narrate the history of life once we are gone? The Stratigraphy Commission may not be formally suspended, but no geoscientist will be there to document the spikes we have left when they finally appear in geologic strata and publicly announce the reality of the Anthropocene.

In 1830, cartoonist-illustrator-geologist Henry De la Beche joked about cyclical notions of history, reappearing species and the end of humans. One of his cartoons shows *Homo sapiens* as fossils, littering the ground, while ichthyosaurs roam free. Will the potential death of humanity be a tragedy or a blessing? Who will have the honour of being the human

Left: Extinction Rebellion activists in red costume attend a mass "die in" in the main hall of the Natural History Museum in London, 22 April, 2019.

endling, the last of their kind? Such questions and concerns are no longer considered science fiction. No doubt they will be increasingly pressing in the near future.[5]

Rethinking the concept of extinction now seems essential. Extinction is not a single event signified by the death of the last organism but a process with a long history and a series of significant consequences. The extinction of the great auk really began in the slaughterhouse of Funk Island, Newfoundland, during the seventeenth century, not in Iceland on 3 June, 1844. Every attempt to prevent an extinction should attend to its advent and aftermath, to connections and context and potential tipping points. Species do not appear with a single event, they renew themselves with every new generation, through the solidarity and mutual care of individuals and collectives, and the shelter and nourishment offered by their habitat.[6]

Awful Changes.
Man found only in a fossil state. —— Reappearance of Ichthyosauri.
"A change came o'er the spirit of my dream". Byron.

A Lecture, — "You will at once perceive," continued Professor Ichthyosaurus," that the skull before us belonged to some of the lower order of animals the teeth are very insignificant the power of the jaws trifling, and altogether it seems wonderful how the creature could have procured food

It is now understood that species can become practically extinct long before the last individuals disappear; many species are on the waiting list for a period, in intensive care in zoos, protected zones or laboratories – in the "death zone", anticipating the inevitable, as the late Australian-American anthropologist Deborah Bird Rose described it.[7] It can also be argued that the extinction process extends beyond the death of the last organisms. Fading away over time, species somehow remain alive in the relations and memories of others that survive.

Mass extinction has become a major international concern. "Extinction Studies" is a growing interdisciplinary field exploring the meaning of extinction, its appearance in different contexts, and future prospects. Extinction Rebellion and all kinds of green movements and institutes address extinction in both the public and private domain. The Center for Biological Diversity, provides "Endangered Species Condoms" as a "fun, unique way to break through the taboo and get people talking about the link between human population growth and the wildlife extinction crisis".[8]

In the Anthropocene, geologic history and human history have fundamentally collapsed into one another, to the extent that one can no longer speak about the planet "itself". While Newton's distinction between unavoidable natural extinction and extinction caused by humans served his time rather well, it now needs radical rethinking, as it fails to inform future attempts to deal with Anthropocenic change. Such a rethink requires an acknowledgement that current environmental hazards are unprecedented, of new kinds and on new scales. The Anthropocene threatens to become a self-inflicted holocaust, a menace to life en masse.

Opposite: *Awful Changes* by Henry De la Beche, with the subtitle "Man found only in fossil state. Reappearance of Ichthyosaurus", 1830.

Above: Endangered Species Condoms by the Center for Biological Diversity.

Ignorance and denial

It is vital for democratic decision-making to establish reasonably accurate information on the state of the planet and the scale of human impact. This is not a simple matter, however, given the complexity of the issues and the multiplicity of accounts: often the public is lost in a jungle of competing political statements and scientific reports. One of the problems of democracy and responsible public policy is plain naivety, partly as a result of the cloud of expert information and deliberate misinformation to which people are exposed. Many political leaders and some experts continue to claim that there's no such thing as global warming and, if there is, that it is not the result of human activities. While the global community of climate experts disagrees on many issues, it has repeatedly confirmed that rising temperatures are partly the result of human activities, driving the planet towards critical "tipping points".

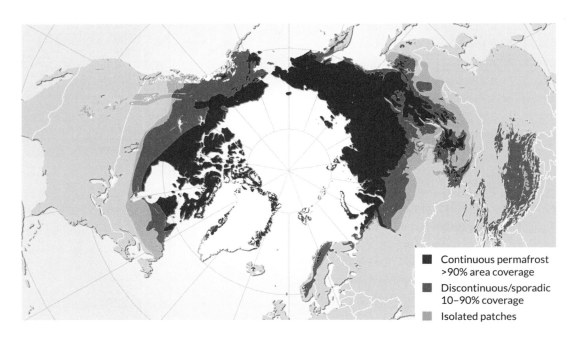

■ Continuous permafrost
>90% area coverage

■ Discontinuous/sporadic
10–90% coverage

■ Isolated patches

British sociologist Linsey McGoey defines "strategic ignorance" as "any actions which mobilise, manufacture or exploit unknowns in a wider environment to avoid liability for earlier actions", adding that strategic ignorance refers "to situations where people create or magnify unknowns in an *offensive* rather than *defensive* way, to generate support for future political initiatives".[1] Strategic ignorance has been endemic in environmental discussions for a long time. Rachel Carson famously battled ignorance with her book *Silent Spring*, published in 1962, which spurred a reversal of US national pesticide policy and helped to inspire a movement that led to the establishment of the US Environmental Protection Agency. A recent republication of Carson's book, along with letters and other writings, reveals the personal courage of its author.[2]

Another telling example of strategic ignorance was the silencing of concerns about the health hazards of tobacco, discussed in Robert N. Proctor's 2012 book *Golden Holocaust*. Proctor exposes the extensive manipulation of public opinion by the tobacco industry, through his analysis of formerly secret industry documents – the Panama Papers, so to speak, of the tobacco industry.[3] Tobacco companies, we now know, conspired to suppress evidence of cancer hazards through funding, fraud and lobbying.

Opposite: Map showing Arctic permafrost.

Above: Heavy industrial smog largely produced by the U.S. Pipe plant in Birmingham, Alabama, 1972.

IGNORANCE AND DENIAL

Is there a resemblance to the debate around climate heating? Recent estimates suggest that the world's largest oil and gas companies spend about $200 million a year on lobbying.[4] Striking lessons emerged in 2019 and 2020, with catastrophic bush fires in Australia that claimed dozens of human lives, ruined thousands of homes and killed more than half a billion animals, pushing some species into extinction and devastating an area the size of Austria. Australian authorities, with Prime Minister Scott Morrison in the lead, continued to deny global heating and human responsibility, treating the fires as isolated events. Direct, respiratory experience of the disaster, as the smoke descended on Sydney, was not sufficient to avert denial.[5] Many people found it symptomatic that the prime minister went on holiday to Hawaii mid-December 2019, without informing the public, as some of the worst firestorms began to rage. Australian media (including News Corp) continued to confound the issue of the fires, attributing them to arsonists.[6] It takes extreme strategic denial, under such circumstances, to suppress recognition of broader climate change and how it is generated.

Australian forest expert Tom Griffiths points out that, while deadly fires have been "burnt into the memories" of many Australian bush dwellers since at least 1851, recent fires are extraordinary in scale and intensity. In recent decades, Australians have come to distinguish between annual "bush fires" and "firestorms" that occur every few decades, punctuating their calendar with the names of specific events, such as the Black Saturday Firestorm of 2009, somewhat in the post-1950s international meteorological fashion of naming hurricanes. Now they are expanding their vocabulary with new terms, including "megafire", to capture the massive merging of bushfires. In Griffiths's words, "individual Black Days have fused into a Savage Summer".[7] Much like nomadic Inuit needed a complex language of "snow" for millennia to adapt to the Arctic terrain, Australians now need a detailed Anthropocenic terminology of fires, with nuances that capture the scale and nature of current events.

From an Australian perspective, the "Age of Fire" would not be a misnomer for this new epoch. The experience of Aboriginal peoples in the fire crisis, it is worthwhile to note, is very different to non-indigenous peoples.[8] For them, settler-colonialism, in the style of the early Anthropocene, is still lived experience. Their strong identification with ancestral land – appropriated, mismanaged and neglected – makes their grief particularly acute.

The official Australian denial may seem puzzling. It has deep roots, however, in national politics and economic interests (particularly around fossil fuels). It is worth noting that in 2014, only five years after the damaging Black Saturday Firestorm, the government of Australia offered the University of Western Australia AUD$4 million to establish a "consensus centre" with, as director, Bjørn Lomborg, the Danish economist who became internationally known as a scholarly denialist in the wake of his bestselling 2011 book *The Skeptical Environmentalist*. Lomborg was formally accused of scientific dishonesty in Denmark and a government committee

Opposite: Evacuation of families in Mallacoota, Australia, on New Year's Eve 2019.

Overleaf: Bush fires outside Bilpin, Australia, 2019.

found his book scientifically dishonest through misrepresentation of scientific facts.[9] Lomborg critiqued the 2012 UN Conference on Environment and Development stating: "Global warming is by no means our main environmental threat". The University of Western Australia accepted the offer to establish a centre for Lomborg, setting off vehement opposition. While the university eventually reversed its decision and rejected the offer, the initial eagerness to spend huge public funds to support Lomborg and the planned centre expressed a consistent failure to address unfolding climate events, such as the recent firestorms.

Interestingly, *The Economist*, which in 2011 featured on its cover the dramatic image of a burning Earth, with the slogan "Welcome to the Anthropocene", came to Lomborg's defence. In an article entitled "Thought control", the magazine suggested that Lomborg's book "measures the 'litany' of environmental alarm that is constantly fed to the public against a range of largely uncontested data about the state of the planet".[10] Lomborg had become a convenient candidate for a "Galileo hypothesis", in which a persecuted loner speaks truth to a regime built on a misleading scientific paradigm. In light of recent floods, weather extremes, bush fires and extinction reports, along with the prognoses of climate science during the last decade or so, this "Galileo hypothesis" does not sound convincing. The Australian government remains committed to an approach of non-interference. According to non-profit thinktank NewClimate Institute, Australia ranks as the worst-performing country on climate policy, in a sample of 57 nations. In the words of Nobel laureate Paul Krugman, "Australia shows us the road to hell".[11]

In the midst of the Australian fires, misinformation spread like an epidemic, including various kinds of unfounded conspiracies, involving eco-terrorists ("greenies") and Chinese billionaires.[12] The channels of misinformation were both social media, including Twitter, and the Rupert Murdoch-owned broadsheet *The Australian*, illustrating a common pattern of climate denial.[13] All of this indicated a clear disinformation campaign, directing attention away from the failures of the Australian authorities and the broad scientific consensus on global heating. These are the times of "fake news".

A recent study by Brown University of millions of tweets shows how computer-generated posts are geared to deny climate science.[14] Robotic postings were responsible for 38 per cent of tweets about "fake science", systematically distorting and misleading online discourse. Sometimes their effects were amplified by people willing to pay for extra visibility. Such activities encourage people to imagine there is more diversity of opinion than there actually is on key points relating to global heating, weakening public support for science. It is tempting to introduce a phrase popularized by Mark Twain on the persuasive power of numbers and three kinds of lies: "lies, damn lies, and statistics".

Climate deniers often insist that their voices, "the other side" in debates about Anthropocenic changes, are fully represented, demanding the right to sit at the table on equal terms. This would be equivalent to inviting tobacco companies as the other side in discussions of public health. In fact, climate deniers systematically avoid fair play, removing actual science from the debate.[15] For instance, in the space of just three years, the US administration has largely excluded science from federal policy-making, halting or disrupting research projects nationwide on a variety of issues, ranging from insects to pollution and climate change. Science is not immune to bias, denial, ignorance and distortion. While a perfect, final scientific account of climate history will never be established, in the long run it is wise to trust the consensus of the scientific community. If we don't attend to the "facts" – messy, shifting and complex as they may be – we may not be able to properly plan and act, until it is too late.

Below: Kangaroos escaping Australian bush fires to save their lives.

Down to Earth

The idea that the Earth is alive may be as old as humankind itself. It is difficult to generalize for different cultures and ages, but overall it seems that until fairly recently, in human terms, the Earth was considered a living entity with no major division between people and their land. Historical evidence from many parts of the world, including the Arctic, Scandinavia and the Amazon, suggests that people belonged to the land, rather than ruling it. In ancient Scandinavia, for instance, people were so indissolubly linked with the land they cultivated that they saw it as an extension of their own nature; social honour was embedded in the land.[1] Humans and the land constituted a single entity, an expanded house, requiring collective maintenance and care. Significantly, the term individual originally meant "indivisible": that which cannot be divided. When humans began to separate themselves from the land, and to see it is an outside entity to be exploited rather than integral to their being, this is when the Anthropocene became first a possibility and then a reality.

During the Renaissance, in the wake of an expansion of commerce during the fourteenth and fifteenth centuries, notions of the unified world were radically rethought. The shift was nicely captured in European art. For Italian

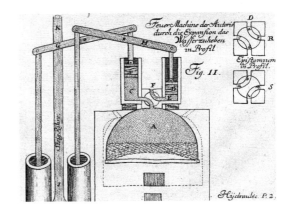

Above: The age of machines. Jacob Leupold's steam engine, 1720.

Opposite above: Prior to perspective. "Lamentation (The Mourning of Christ)". Giotto di Bondone.

Opposite below: The arrival of perspective. "Eventos de la vida de Moisés". Sandro Botticelli.

painters, trained in the static and holistic world of ancient Greek philosophy and the medieval church, the canvas was a two-dimensional space for the glorification of godly designs and beings. By the end of the Renaissance, in contrast, the art of painting focused on human activities and their place in nature and history. Renaissance painters were rewarded for their efforts with a spectacular artistic success, the laws of perspective (*perspectiva*, or "seeing through"). Within a brief period, nature became a quantifiable,

DOWN TO EARTH

three-dimensional universe appropriated by humans. This "anthropocracy", to use the term of art critic Erwin Panofsky, represented a radical departure.[2] Now humans imagined they could separate themselves from the world, watching it at a distance, a precursor to the Anthropocene. The laws of perspective were a powerful and productive innovation, paving the way for modern science. But they came at a cost.

The concept of oneness with the land nevertheless survived. Many twentieth-century anthropological accounts testify to concerns with the unity and vitality of the land and the people. One account describes the "economy of livelihood" in rural Colombia, the ways in which human bodies are embedded in the land and the manner in which people "care for the base", as if the land were an extension of themselves, maintaining its *forza* ("force" or "strength") in order to ensure sustainability, to use modern environmental jargon.[3] For Colombian peasants, in other words, social relations and natural, earthly relations are inextricably linked.

In recent decades, the duality of material Earth and the social life of humans has increasingly been challenged. In particular, the Anthropocene – or rather the *recognition* of human impact and dependence on the planet – has called for a realignment of the social and the geological. In the humanities and social sciences, the term "social" has long served as an unquestioned shorthand for human-to-human relations. Today, many people are extending the notion of sociality not only to other species, plants and animals, but also to material things, including rocks and mountains.

For many of us, geological things and forms don't seem to participate in social life according to common criteria: they don't seem to have interior lives, they don't strive, they don't metabolize, they don't reproduce or seem to respond. However, an increasing number of scholars in a variety of disciplines, are offering new terms with which to think about Earth-human relations, taking the material dimension seriously. Thus, geoscientists have become more open to the possibility of socializing the material world, as witnessed with the geological discourse around the Anthropocene, which emphasizes the conflation of the human and the geologic. Some humanities scholars, social scientists and artists likewise speak of a radical turn

in their fields, from the reverse perspective, with a materialization of the social world.

One example is the notion of "geological intimacy", which draws attention to the ways in which humans feel attached to material nature.[4] The term was launched by American artist Ilana Halperin. In 2003, she issued an open invitation to a thirtieth birthday party for herself and Eldfell, a volcano in the Westman Islands, Iceland, which only rose out of the sea in 1973: "You may be asking yourself if I am serious about this invite and the answer is absolutely yes, as, let's face it, you and a landmass only turn 30 at (almost) exactly the same time once! All the best and see you at the crater!" It was a windy day and the candles struggled to stay alight up on the mountain, but the birthday cake tasted good. Some months later, Halperin's birthday happening found a manifestation in her art, and her works about her intimacy with the Earth have been shown widely. They are lyrical pieces that remind people of the indispensable comradeship of the Earth, of global connections and life's coincidences.

Another person socializing with the material world is American political scientist Jane Bennett, who aims "to articulate a vibrant materiality that runs alongside and inside humans to see how analyses of political events might change if we gave the force of things more due".[5] After all, humans are made of the same elements as our planet (hydrogen, carbon, sulphur, etc.), cannot survive without them and sometimes can't help noticing them (think of gallstones which fuse rocks and human bodies). As Russian chemist Vladimir I. Vernadsky put it, "the material of Earth's crust has been packaged into myriad moving beings whose reproduction and growth build and break down matter on a global scale... *We are walking, talking minerals*".[6]

Opposite: Eldfell volcano, Westman Islands, Iceland.

Above: Chichibu Chinsekikan, hall of curious rocks, a museum near Tokyo, Japan.

The notion of geosociality seeks to sum up these ideas, evident in practically all kinds of scholarship as well as the arts.[7] One of the important themes on the modern agenda, in both research and the arts, is to explore changes in human engagement with the planet in the wake of extreme weather, glacier decline and other Anthropocenic developments. Can we relate to and identify with rocks, mountains, rivers and glaciers, just as we relate to fellow humans and the animal kingdom? In fact, we often do, in a manner that echoes medieval and more ancient notions of belonging to the land.

In 2017, New Zealand granted a sacred mountain, Mount Taranaki on the North Island, the same legal rights as a person. This was done in response to demands by Māori tribes for respect for their place and identity, an apology for broken treaties and promises, and to protect against the growing harms of tourism. For the Māori, Mount Taranaki, a well-formed volcano that last erupted in 1775, is close kin, a member of the family.

Geosocialities are intertwinings of bodies and biographies with Earth, demanding we pay attention to how geology matters in different ways to different cultures and people in different locales. One striking case of geosociality is a museum near Tokyo, Japan: Chichibu Chinsekikan or the Hall of Curious Rocks presents 900 rocks resembling human faces.[8]

Geologic intimacy and geosociality should not be treated as either trivial or exotic, mere things of the past. Rather, they should be held up as beacons of hope – if only more societies were willing to identify more closely with nature, would the threat of the climate crisis be so grave?

Right: Mount Taranaki, New Zealand.

DOWN TO EARTH

TWENTY-TWO

The case of freezing lava

"No physicist," claimed nineteenth-century American environmentalist George Perkins Marsh, "has supposed that man can avert the eruption of a volcano or diminish the quantity of melted rock which it pours out of the bowels of the earth."[1] Although he listed examples of efforts focusing on diverting lava flows, some dating back to the seventeenth century, these were limited in scope. The idea of the Anthropocene had not arrived, but wouldn't freezing lava be

a test case of large-scale human impact on the Earth itself? A surprise opportunity came in 1973 in the Westman Islands, Iceland. This was a successful effort to inscribe the human in rock during an eruption. More than any other event, perhaps, this case challenges Marsh's notion of the "impotence" of humans when confronted with lava flows.[2]

The main settlement on the Westman Islands, south of the mainland, had about 5,000 inhabitants at the beginning of 1973. The natural harbour provided safe moorings for one of the biggest fishing industries in the country, close to rich fishing grounds. On 23 January, the volcano Eldfell, on the outskirts of town, erupted. In the middle of the night, the ground suddenly opened with a roaring sound as the volcano sent glowing lava into the sky and down towards the harbour.

While the Westman Islanders were taken by surprise when the eruption began, at close to two in the morning, they acted swiftly when awoken. In a few hours, most of the people on the island were safely transported over rough seas to the nearest mainland harbour. The next concerns were to prevent houses from collapsing under the ash that fell like heavy rain and, above all, to try to redirect the lava flow to avoid the destruction of the harbour.

A somewhat eccentric physics professor, Þorbjörn Sigurgeirsson, came up with the idea

Opposite: One of the early photographs of the Westman Islands eruption.

Below: The fishing harbour and the community of the Westman Islands under threat during the 1973 eruption.

Overleaf: Four time-lapse photos taken during the Westman Islands eruption.

THE CASE OF FREEZING LAVA

of cooling the lava. He suggested trying to pump water on the advancing lava front in order to slow it down or halt it, beginning with the local fire brigade's truck. Most Icelanders thought that this was an absurd proposal. The force of Mother Nature, it was argued, could not be tamed by "having a pee" on the edge of the glowing lava.

During the ensuing weeks a complicated story unfolded, sometimes called the "battle with lava". It seemed that the initial cooling had some impact, but Sigurgeirsson and many others reasoned that the available pumps were not powerful enough to have much effect. Following a series of explosions in the crater, massive lava flows headed north towards the harbour and west towards the town, with many houses burning or collapsing. Sigurgeirsson managed to convince the authorities to arrange more effective water pumping on to the lava – on an unprecedented scale. It was a race against time. Through an agreement with United States authorities, about 40 massive pumps were quickly shipped to the Westman Islands.

A professor of mechanical engineering, Valdimar K. Jónsson, was hired to organize the

Below: Retreat from the electrical power station.
Opposite: Cooling lava on the edge of town.

pumping while Sigurgeirsson would decide on the strategy on a daily basis, depending on the movements of the lava. The water clearly had the effect of freezing the lava on the edges, but the pressure of the flow repeatedly threatened to break down the walls that had just been created by cooling. Later on, pumping crews would take the pipes on to the lava, approaching the crater itself in advancing steps, directing the water flow on to the lava with the aid of a bulldozer and a crane. This was risky for the crew at the front as their boots might burn, and they might get trapped. The design, coordination and deployment of the pipes

(first steel and aluminium, and later plastic) was a major engineering feat. After weeks of intensive pumping and constant movement of pipes and people back and forth, the lava was redirected eastward, away from the harbour and into the ocean. According to the United States Geological Survey, it was the greatest effort ever attempted to control lava flows during the course of an eruption. Eventually, the volcano calmed down. On 3 July 1973, Sigurgeirsson and colleagues descended into the crater, and declared the eruption over, sparking scenes of jubilation. The Westman Islanders were able to return to their homes.

Human agency, then, expressed via the pumping of enormous quantities of sea water on to targeted areas of flowing lava, significantly slowed down and solidified the fast-moving mass, avoiding the destruction of an important fishing harbour and a number of houses. The volume of water pumped was the equivalent, said writer John McPhee when documenting events on the Westman Islands, "of turning Niagara Falls onto the island for half an hour".[3] While humans apply their labour to solid rock in the course of many other activities, especially in mining and tunnel operations, this was significantly different. In successfully cooling lava, humans had inscribed their influence into rock during the process of its formation, drafting a new chapter in history. The result was a peculiar form of water-hardened stone. "Among the natural patterns of lava flows, it was utterly anomalous. In a very certain sense, it was man-made", noted McPhee. "After the human contribution passed a level higher than trifling, the evolution of the new landscape could in no pure sense be natural. The event had lost its status as a simple act of God."[4]

Today, Westman Islanders commemorate the eruption of Eldfell in many ways, partly as a way to deal with the trauma of the recent past. A collective celebration of the "closing" of the eruption (*goslok*) takes place every July. Recently, a museum named Eldheimar ("Worlds of Fire") was established to present imagery, photos and

2. júl 1973
Foto: G. S.

film from the time of the eruption. The museum, sometimes referred to as the "Pompeii of the North", was erected around a house that was half-destroyed by ash, demonstrating how the family who lived there had left in a hurry without warning, leaving everything just as it was in the course of everyday life.

Volcanoes have immense power and agency, usually unaffected by humans or other living beings. While the volcano was obviously the central agent in the story about the cooling of lava in the Westman Islands, transferring glowing magma from the depths of the Earth to its surface and distributing it in all directions, there were other agents as well. The hydraulics of the pumps and the pipes in the harbour, and on the boiling and moving lava – an ingenious assembly fuelled by diesel and petrol (again earthly, geological stuff) – played a central role. In the process of cooling lava, saving a harbour and rescuing a community, the pumps facilitated many other acts as well: organizing collaborations and crafting narratives, often with a good deal of irony and humour about fire and earth, and about human engagement with volcanic activity.

Opposite: The eastern part of the town of the Westman Islands in the middle of eruption.

Above: In the bottom of the crater of Eldfell, Westman Islands, declaring the eruption over. First from right: physicist Þorbjörn Sigurgeirsson, architect of the cooling operation.

Is There Hope?

Missed opportunities

Keeping in mind all the evidence accumulated regarding the damaged planet and the challenges ahead, the next steps needed should be readily apparent. Yet they are not, partly because of the climate of mistrust in science. How did this come about? Two historical cases help to set the stage: one, the case of the ozone layer, a success story that gives grounds for optimism, and another, the sad story of so-called Climategate, which planted some of the seeds of denial and mistrust. The global environmental crisis presents unprecedented challenges to human cognition and discourse. Some of these challenges relate to the limits of direct perception and our inevitable reliance on virtual representations. American environmental historian William Cronon remarked in 1996:

... some of the most dramatic environmental problems we appear to be facing ... exist mainly as simulated representations in complex computer models of natural systems. Our awareness of the ozone hole over the Antarctic, for instance, depends very much on the ability of machines to process large amounts of data to produce maps of atmospheric phenomena that we ourselves could never witness at first hand. No one has ever seen the ozone hole. However real the problem may be, our knowledge of it cannot help being virtual.[1]

The ozone hole is no longer on the public radar, but it dominated environmental discussions in the 1970s and 1980s, becoming a worldwide sensation.

The ozone layer, or ozone shield, is a region of Earth's stratosphere, which is the second major layer of the atmosphere. Containing high levels of ozone (O_3), it absorbs most of the Sun's dangerous ultraviolet radiation. Without it, cancer deaths and crop failures would be endemic on Earth. Ozone was discovered in 1893 by German chemist Christian Friedrich Schönbein, who had noticed a bad smell as he experimented with electrolysis of water. He named the unknown substance he produced "ozone", from the Greek word *ozein*, or smell. The ozone layer, at an altitude twice the height of Mount Everest, was discovered in 1913 by French physicist Charles Fabry. For decades it was assumed the protective layer must be permanent, beyond human influence and thus nothing to worry about.

Two scientific papers published in 1974 changed this perspective. The authors, American chemists Frank Sherwood Rowland and Mario Molina, showed that chlorine that was vented into the atmosphere was destroying the ozone layer – a discovery for which they won the Nobel Prize in 1995, along with Dutch

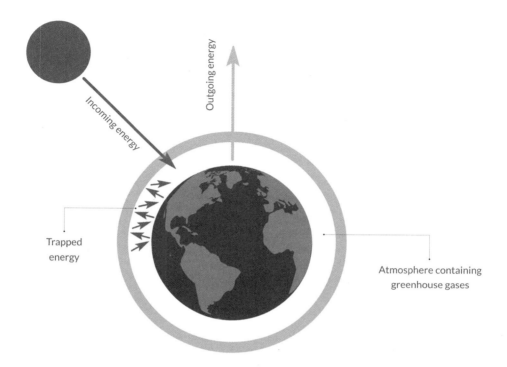

chemist Paul Crutzen, who later popularized the term Anthropocene.[2] The discovery of the ozone hole was a shock. Soon the culprit was identified. Chlorofluorocarbon molecules (CFCs), from spray cans, refrigerators and other related human products, were entering the stratosphere, destroying the ozone layer. These were controversial findings, but evidence presented between 1977 and 1984 clearly demonstrated that the concentration of ozone above Antarctica had plunged more than 40 per cent below the 1960 baseline, exposing humans and other living beings to progressively higher levels of dangerous ultraviolet light, which tears apart organic molecules.

By the late 1980s, there was broad scientific and public agreement about the effects of CFCs and their serious environmental implications if the global community failed to respond.[3] Producers of spray cans and refrigerators offered no major opposition. The United Nations Environmental Program hosted the international discussions that were needed, leading to forty-nine nations signing the Montreal Protocol of 1987, which mandated 50 per cent reduction in the production and consumption of CFCs by 2000. Later, stricter measures were taken to phase out CFCs entirely. The language of tipping points

Above: Ozone depletion and the greenhouse effect.

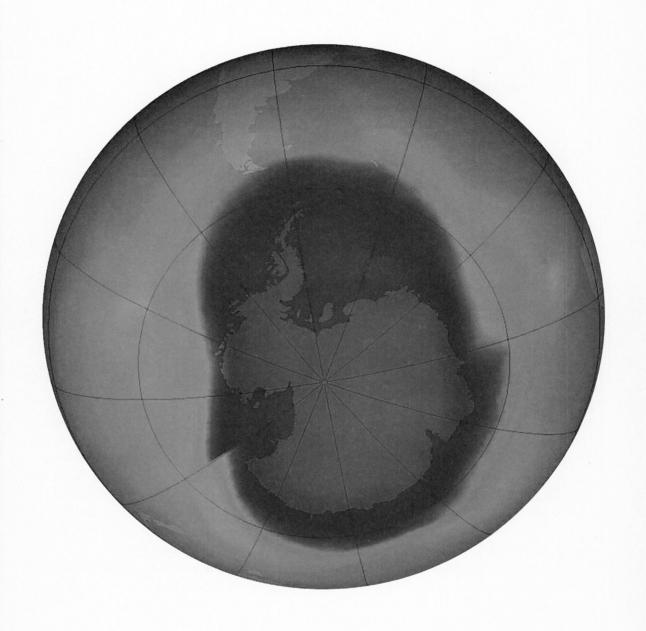

172

had not arrived, but the global community effectively brought the ozone problem under control, to avoid irreparable damage. Practically forgotten by now, this was a story of progress and success.

One would have thought that the issue of global warming might have taken a similar course, with agreement on basic facts, key problems, and effective measures to take. Six years before the ratification of the Montreal Protocol, James E. Hansen and colleagues had pointed out the progression and dangers of global heating. Yet the issue of global heating was quickly derailed, with lasting consequences for climate-change discussions worldwide. One of the drivers of that derailment was Climategate, an affair which, like the disappearing ozone layer, is practically forgotten – in spite of a still formidable presence on Google and a rich associated literature.[4]

Climategate began in 2009 when somebody accessed the computers of an important laboratory of climate science, the Climatic Research Unit (CRU) of the University of East Anglia in Norwich. Before long, more than 1,000 emails and about 3,000 climate data documents had been extracted and uploaded to a server in Tomsk, Russia, eventually circulating throughout the world via email and the internet. Critics of the climate scientists argued, on the basis of their readings of the material, that the thesis of global warming was a "hoax", claiming that the leak showed manipulation of data, silencing of dissent and a lack of "transparency".

Much of the discussion focused on extrapolations of global temperatures, partly based on "proxy" climate data, in the form of tree rings. The so-called "hockey stick graph", showing the acceleration of temperatures in recent years, drawn by Michael E. Manning, director of the CRU, became a contested icon. Critics, some of whom called themselves "deniers", argued that Manning and his colleagues had fiddled with the data, playing tricks on their readers to make their stories more convincing. Climategate rang some bells. Perhaps a more appropriate name nowadays, at the time of WikiLeaks and the Panama Papers, would be the Climate Papers.

Following the disclosure of the documents, there was a heated public discussion in Britain and internationally, with extensive media coverage, responses from the climate scientists involved and formal public enquiries. There was much at stake. The timing of the hacking of the CRU computers was probably not accidental. It coincided with the Copenhagen Climate Summit and helped to undermine international efforts to address global heating. In the ensuing years, the damaged hockey stick had a second life in the growing circle of climate deniers, sometimes in scientific think tanks funded by oil billionaires, including the Cato Institute in Washington, D.C., which advocates for individual freedoms against the excesses of government control. Lost in the storm was the successful battle to repair the thinning ozone layer.

New forms of openness, collaboration and exposure are now emerging, as well as new forms of participation and engagement (such as citizen science, user-led innovation, participatory sensing and crowdsourcing).

Opposite: The largest Antarctic ozone hole recorded, September 2006.

With hindsight, a decade after the disclosure of the Climate Papers, the problems associated with proxy data were overblown and the critics' idea of pure science seems naive and outdated. Likewise, the scientists involved operated with old-fashioned notions of the intellectual ivory tower, barely able to address public environmental concerns.

We are back to the notion of the aquarium, which divides participants from spectators, laypeople from experts, nature from society. While Climategate was a limited affair in the broad scheme of things, it was nevertheless an important example of the problems of mistrust between science and the larger community, the dangers of fake news, and the growing role of the media. It is vital for humanity to learn from the past, and current records as well as recent evidence clearly demonstrate accelerating heating and its association with weather extremes and massive floods.

Above: The Cato Institute, Washington, D.C.

Opposite: Members of the Select Committee on Energy Independence and Global Warming and the House Republican American Energy Solutions Group attend a press conference on the Climategate Scandal, 2009.

MISSED OPPORTUNITIES

TWENTY-FOUR

· ·

Engineering Earth

It may be tempting to see the cooling of the lava on the Westman Islands as an isolated and innocent battle of "man against nature", achieved through the application of scientific expertise and a powerful pumping system. In fact, it was intimately connected to larger developments, notably the US military establishment, nuclear development and the Cold War. For one thing, the American pumps had military connections, as they had previously been used in Cold War operations, delivering fuel on enemy soil. Also, the architect of the cooling operation had been trained with Niels Bohr and his colleagues in Copenhagen, in a leading laboratory partly focused on nuclear research.[1] The "battle with lava" during the eruption of 1973, in fact, was partly informed by military concerns with "planetary" scale earthquakes and unusual weather[2], precursors to Anthropocenic concerns with damaging climate change and the possibility of averting it by engineering means.

Recently, engineering has played an increasing role in discussions of the Anthropocene, offering ideas and solutions on practically everything climatic and planetary, from minor technological tricks to large-scale interventions to halt the exponential growth of atmospheric CO_2 during the last decades.[3] Humanity's inventiveness and engineering have obviously achieved a great deal in many fields; we have captured solar energy, harnessed nuclear energy. At the same time, we have created escalating environmental problems which have no obvious solution.

For decades, the practical measures to deal with global heating have been adaptation and mitigation, based on the assumption that there is ample time to experiment with lowering CO_2 to secure levels for the future of the planet. Some of the key measures involve setting carbon taxes and emission quotas, expanding forests and advocating renewable energy and non-fossil sources, including solar power. Mainstream economists have long assumed that it would take a whole century for CO_2 levels to double, and that this would allow time for market measures to work.[4] Machinery, buildings and flood control systems, it was believed, could be modified faster than the pace of global warming.

Opposite: Spillway from a flood-control dam.

Above: Stratospheric aerosol injection, a process that could be used to limit solar radiation by causing a global dimming effect.

Now it has become clear that CO_2 levels have accelerated much faster than earlier assumed. Also, while the "invisible hand" of the market and the financial world may fix some problems (e.g. putting a value on plastic bottles affects consumption and disposal), overall it is only adding to the chaos. At the same time, there is a lack of international commitment, and political backlash in many contexts, with a return to climate denial and fossil-fuel exploitation (in contexts as culturally and politically diverse as the United States and Norway). As a result, there is a growing sense of urgency, if not panic. For many critics, it is time to work seriously on two fronts: geological engineering (fixing the carbon problem through inventive chemistry and technology) and radical restructuring of society, with an overhaul of the forces that favour capital over life, private interests over collective ones – some kind of Green New Deal.[5]

"Geological engineering" is a term historically associated with mining and the extraction of fossil fuels; now it has expanded in the other direction, into space, with so-called "climate engineering": changing the way sunlight reaches the planet, to limit the energy budget of the Earth. This might involve screening sunlight with space mirrors, stratospheric aerosol injection, reflective balloons or stimulation of cloud condensation. Science fiction has come down to Earth. Such measures may be costly, compared to many attempts to modify the scale and effects of global heating, but relative to the harms of heating, they may well be a bargain. Also, they may work more quickly than other measures. However, they pose numerous risks

Right: Floods in Manggarai, Jakarta, Indonesia, 2020.

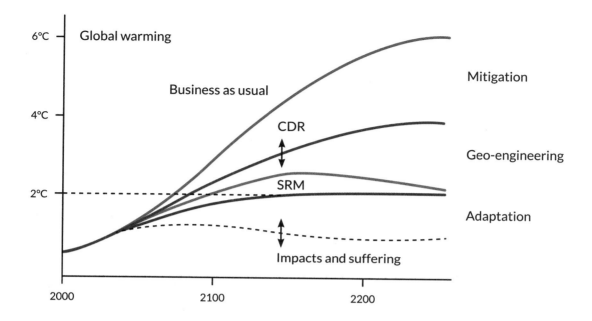

Global warming

6°C

Business as usual

Mitigation

4°C

CDR

Geo-engineering

SRM

2°C

Adaptation

Impacts and suffering

2000   2100   2200

and problems. "Re-engineering" the Earth on a large scale means diving into new territory, with extreme risks of potential misfires and drastic side effects, both human and environmental.

In her important book *After Engineering*, American environmental researcher Holly Jean Buck emphasizes the value of moving beyond "climate futures ... described in terms of mathematical pathways or scenarios, behind which are traditions of men gaming out possible futures". Buck's own strategy is to introduce fiction into mathematics and theory, "to make the future less empty, to populate it with embodied lives and emotions".[6] Another significant genre is that of autobiography, exploring real-life experience and life histories.[7] Robert Macfarlane's *Underland: A Deep Time Journey* is a prime example of this relational storytelling. Still another avenue is genuine Anthropocene fiction in the fashion of Max Frisch's novel *Man*

in the Holocene mentioned earlier, which possibly launched a new genre, emphasizing the errors of assuming that sea levels are "fixed for all time" during "days like these".[8]

The visual arts are vital too, appealing to public sensibilities through all kinds of media. One striking environmental work is *Ice Watch* by Copenhagen-based artist Olafur Eliasson and geologist Minik Rosing, exhibited in 2015, first in Copenhagen and later at Place du Panthéon in Paris. The installation contained several heavy chunks of ice from Greenland. Eliasson is known for his engaging environmental works (including *The Weather Project* at Tate Modern in London in 2003) while Rosing is recognized for his groundbreaking study of photosynthesis

Opposite: Sunlight screening with space mirrors.

Above: Using solar geoengineering to "shave the peak" off a temperature overshot. The so-called "napkin diagram" of John Shepard, 2010.

in the Greenland sea bed, which pushed the beginnings of life 200 million years backwards.[9] Apparently, one of the points of their installation was to explore relationships between people, environmental data and emotions – and, indeed, observers of the melting ice were touched and impressed. All of these artistic genres, textual, conceptual and visual, no doubt, will play an important role in future attempts to circumscribe and fathom our age and to imagine our futures.

The Anthropocene, it is often pointed out, is a highly anthropocentric construct, tainted by human bias, an engineering ethos and a passion for control. The comeback of Anthropos, in the modern (or post-modern) era, represents an ironic conceptual feat, returning humans to the driver's seat in history, from which they were recently expelled with the loss of faith in science and progress. After all, there are good grounds for speaking of collective or distributed, more-than-human agency: to emphasize that the Anthropocene is not the result of *Homo sapiens* acting in isolation. It is only made possible through a diverse network of biological, technological, cultural, organic and geological entities. Nevertheless, humans are both important agents and often acutely aware of their doings.

One of the conceptual problems of Anthropocene talk is that observers of the environmental crisis, and the languages available to them, are necessarily embedded in the world they observe. To what extent, for instance, are the metaphors we now live by – including those of tipping points and windows of opportunity – reliable, effective and innocent language? The question is particularly pertinent in the Anthropocene. How can we meaningfully deal with the current crisis if we are becoming just as Earth-bound as tectonic plates, if our hands and bodies are literally fossilized, engraved in geologic strata along with plastic and chicken bones?[10] What does the Anthropocenic conflation of *geos* and culture entail for freedom, objectivity and responsibility?

The Anthropocene does not just imply conflation of the natural and the social, Earth and humanity, but it also suggests a radical change in perspective and action in terms of human awareness and responsibility, a "new human condition", to paraphrase German philosopher Hannah Arendt. Arendt wrote about social developments in the wake of the Second World War, including human alienation from nature and the changing character of politics, science and freedom, in her influential book *The Human Condition*.[11] To what extent, one may wonder, has the human condition changed during the Anthropocene?[12] Often, significant change is only understood after the fact, decades or centuries after a new age or phenomenon establishes itself. Hindsight is not a choice, in this case. Creative fiction and the visual arts are indispensable, illuminating our lives, our circumstances and prospects, along with the humanities and social sciences.

Opposite: *Ice Watch*, an installation by Olafur Eliasson and Minik Rosing. Place du Panthéon, Paris, France, 2015.

....

TWENTY-FIVE

Fixing carbon and buying time

In the spring of 2009, the ambitious Iceland Deep Drilling Project came to a sudden halt. It was looking for deep sources of geothermal energy at the Krafla volcano. When drilling progressed to a depth of 2066 metres, the machinery refused to continue.[1] It turned out that the drill had, unexpectedly, reached into the mixture of molten and semi-molten rock between the Earth's surface and above the mantle which wraps around the planet's core. The drill had ventured into a magma reservoir – a rare occurrence despite thousands of geothermal drilling projects worldwide. Drilling resumed for a while with the pumping of cold water into the well, but eventually the well was shut down after a valve failure.

Keeping in mind the threat posed by major magma extrusions (i.e. volcanoes) throughout history, some of the experts and technicians at Krafla no doubt wondered if drilling into magma would prove catastrophic. Hadn't this encounter with magma violated the long-standing barrier between the "dead" geologic interior of the planet and its living, cultured surface? Would the stunning opportunity to "touch" magma –

to study it before it cooled and lost its fleeting, magmatic qualities in some lava flow – open a Pandora's box, perhaps disrupting gravity, bending time or even signalling the end of it?[2]

In fact, a number of drilling projects in different parts of the world are currently *buying* time by pushing carbon dioxide into the Earth's crust. Such projects, innovative and important examples of modern geoengineering, take various forms, relying on different kinds of technology and chemical approaches. One of the early subterranean projects is based at the Sleipner gas field west of Stavanger, Norway. Between 1996 and 2014, the Sleipner project captured one million tons of CO_2 annually, storing it more than 700 metres below the ocean floor in permeable sandstone via physical trapping. Mineral trapping is expected to take thousands of years at least at such a site due to low reactivity and lack of the chemical elements needed for mineralization, the safest storing mechanism. Physical trapping requires a lot

Below: The site of the Iceland Deep Drilling Project in the north of the country.

FIXING CARBON AND BUYING TIME

of space and may not be secure, obviously an important disadvantage.

Other projects have managed to achieve fast and safe mineralization through injection. A pioneer in the field is project CarbFix based in southwestern Iceland, initiated in 2006 and funded by the European Union and the US Department of Energy.[3] After several years of preparation, lots of international diplomacy, academic collaboration and cooperation with an Icelandic geothermal energy project (at the Hellisheiði geothermal plant), a pilot carbon injection was carried out in January 2012. CO_2 was released as small gas bubbles into downflowing water within an injection well, the bubbles dissolving in the water. The CO_2-charged water accelerated metal release from basal rock at a depth of 500 to 800 metres, forming solid carbonated minerals within two years with more than 95 per cent of the injected CO_2 becoming mineralized. Scaling up two years after the first CO_2 CarbFix injection, in a hotter and deeper reservoir, more than 50 per cent of the carbon mineralized within months.[4]

These were striking results, given that early predictions of the time required for mineralization were that it might take years, decades or even centuries. As Sigurður R. Gíslason and Eric H. Oelkers, two of the project's main architects, put it: "Once stored as a mineral, the CO_2 is immobilised for geological time scales".[5] In the original phase, CarbFix injected 175 tons of pure CO_2 into porous basalt rock, then 73 tons of a mixture of CO_2 and other gases from the Hellisheiði geothermal plant. In its second phase, CarbFix2, the project drew upon over 100

Right: The carbon storage site of the Sleipner gas field west of Stavanger, Norway.

FIXING CARBON AND BUYING TIME

geothermal wells of varying depths (down to 3,300 metres) which facilitated careful monitoring of the fluids injected into the ground using tracers and isotopes, and the process of mineralization.[6] In the early phases CarbFix generated seismic activity, much like fracking in some parts of the world. Early on, two magnitude 4 earthquakes occurred, but soon after, seismic effects were minimized (in 2018, only one magnitude 2 quake occurred).[7] A project similar to CarbFix is the Big Sky Carbon Sequestration Partnership in Wallula, Washington, United States, which began in 2013.[8]

Most of the projects involving the capture of carbon are in an experimental phase, balancing costs, carbon footprint, technology, resources, political concerns and human skills. Some, however, are close to being developed to industrial scale. While CO_2 mixes rapidly in air and travels well, resulting in uniform concentrations around the globe, carbon capture has proven most effective when carried out close to the point sources, such as industrial plants and power stations. Some projects still in the early stages of development seek to capture CO_2 directly from the atmosphere, in a similar fashion to the recent experiments with air conditioning, channelling air into filtering units to selectively remove CO_2. Two examples are CarbFix2, an extension of earlier experiments at the geothermal site in Iceland, and the Solid Carbon Project in British Columbia, Canada. There are even exploratory projects seeking to combine technology for capturing CO_2 directly from the atmosphere with offshore sequestration of CO_2 in submarine basalt rock strata, for carbon mineralization.[9]

All of these projects are genuinely Anthropocenic in the sense that, once again, humans are seeking, through drilling and pumping, to make a lasting impact on the planet, refashioning its underworld and helping to prolong terrestrial life as we know it. All of them seek to buy time by scaling down the impact of fossil fuels before the planet reaches

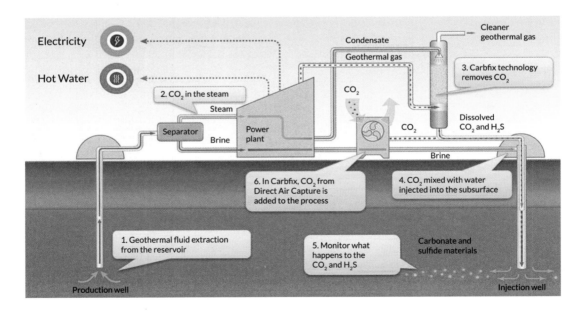

projected tipping points, mineralizing CO_2 over a surprisingly short period of time. The resulting spikes in the strata will be visible for millennia, in the Earth's crust and, hopefully, in the biosphere. Carbon continues its long and winding journey around the carbon cycle, sometimes acting as a key to life and sometimes becoming a deadly threat.

Such projects are likely to expand and multiply in coming years and decades, mainly because capturing CO_2 and storing it safely will become imperative, as emission quotas bite, carbon pricing becomes more expensive, and temperatures continue to increase. The amount of CO_2 that is captured at current rates is minimal compared to the 190 gigatons that would have to be stored in order to achieve the goal of limiting global heating to a rise of 1.5 degrees Celsius. Nonetheless, such projects clearly provide an important avenue for

development in the Anthropocene.[10] But they will not be applicable everywhere, as they are costly and dependent on a host of local factors, including specific rock formations and ample water supplies. In the long run, carbon capture and storage may become less important, assuming that the global community manages to reach the goal of minimal or zero emissions. Some uses of fossil fuels, on the other hand, can be expected to continue into the distant future, while capturing CO_2 – by whatever means – is likely to remain critical.[11]

Opposite: Schematic presentation of the combination of the Geothermal Power Plant and CarbFix technology at the Hellisheiði site in southwestern Iceland.

Above: Core from injection site showing CO_2-bearing carbonate minerals within basaltic host rock; CarbFix, Iceland.

Overleaf: The CarbFix site and Geothermal Plant on Hellisheiði in southwestern Iceland.

Rebellion

Environmentalism – the quest for a viable future – was partly born with the works of nineteenth-century North American writers, among them Ralph Waldo Emerson, Emily Dickinson and Henry David Thoreau. Emphasizing human experience of nature and human responsibility for preserving the wilderness, Thoreau's testimony in his *Walden: Or, Life in the Woods*, about living around Walden Pond in Concord, Massachusetts, remains one of the most influential environmental statements of all time. As one of Thoreau's biographers has written, "the railroad whistle across Walden Pond sounded the death knell of an old world and the birth of something new … when fossil fuels put global economies into hyperdrive, the Anthropocene".[1] Thoreau's stay in the woods was not only an opportunity to write in solitude, it was a public protest for those who bothered to pay attention, a call for action akin to the civil disobedience of current environmentalists, for whom planet Earth mirrors the small world of Walden Pond and the woods surrounding it.[2]

Environmental political parties and movements are common in many contexts, but a recent international movement, Extinction Rebellion (XR, *@ExtinctionRebellion*), has struck a new chord, driven mostly by young people who know that they will inherit the problems of the

Anthropocene, determined to push for radical change with organized nonviolent public protests. The movement was founded in 2018 by two British activists, Gail Bradbrook and Roger Hallam, who drafted a strategy for mass civil disobedience focusing on climate disaster and disappearing species. Their activities and manifesto are documented on their website and their handbook *This is Not a Drill: An Extinction Rebellion Handbook*.[3] During the last two years, mass protests under the banner of Extinction Rebellion have shaken many cities.[4] Invariably, there are demands for immediate action on the climate front, often followed by the slogan "There is no Planet B!"

The movement consists of organized cells in numerous towns and cities throughout the world, linked by the internet and social media, which facilitate swift action and efficient organizing. It is fitting that Cambridge, UK, the base of zoologist Alfred Newton, who pioneered extinction studies in the nineteenth century, has a strong Extinction cell. Its core consists of young people from different walks of life, angered by public denial and governmental failure, who share serious concerns about their future prospects. They organize locals

Opposite above and below: Extinction Rebellion protest in London's Parliament Square, 2019.

and neighbours, setting agendas and taking part in demonstrations demanding radical change. Few of them have participated in any kind of politics before, but now they are suddenly captivated, gathering in huge crowds and organizing events that draw attention to their cause. Some liken their unexpected radicalism to being struck by lightning. Many of the older people who join them at meetings are tearful, out of fear for the future of their grandchildren.[5] In the autumn of 2019, the Cambridge team participated in a series of London events, joining tens of thousands of people from other parts of England, Scotland and Wales, demanding the British Government publicly acknowledge the climate emergency. Events throughout the world have captured public attention, disrupting traffic and routines.

In the Anthropocene, democratic participation and critical thinking about the environment seems particularly important among even the very young. Many cities, including Berlin, Birmingham, Seattle and Tokyo, have been trying to cultivate civil participation among young children, creating safe spaces for them in kindergarten and primary school, where they can experiment with architecture and social life. In the process, children are encouraged to craft their own mock shops, roles, and institutes, as they prepare for meaningful politics, active participation and a better future.[6] Japanese kindergarten architect Takaharu Tezuka and his wife Yui Tezuka have played an important role with their groundbreaking perspectives and architecture, encouraging students to design their own learning environment through collaborative apprenticeship.

Left: Extinction Rebellion protest in London's Parliament Square, 2019.

Perhaps youthful activism of this kind could serve as a model for climate action. It was, however, a young Swedish girl, Greta Thunberg, who almost single-handedly ignited a global youth movement of "climate strikes", as she put it, with her demonstrations.[7]

Greta launched her first strike at age 16, outside the Swedish Parliament in Stockholm, in August 2018. She had been depressed since learning about climate heating and rising sea levels at school, five years earlier. Now she was back on track, regaining her energy. On her first strike, she was alone, but soon others joined her on her "Fridays for Future" events. By September 2019, climate strikes had spread throughout the world with millions of people participating, under the joint banner of Greta and Extinction Rebellion.

Standing firm on her principle of limiting her carbon footprint and avoiding long flights, she sailed on a small boat to UN meetings in New York City and Madrid, where she joined global leaders and climate protesters.

The so-called "Greta Effect" has been immense. Margaret Atwood compared Greta to Joan of Arc. In 2019, she was declared Person of the Year by the *Time* magazine. Most importantly, she has become a symbol for global rebellion, an idol for a whole generation, galvanizing the public with her straight talk and her insistence that people respond to climate urgency by both changing their lifestyle and by working towards radical, structural change. More than any statesperson,

academic, or scientist – and despite her modesty and self-effacing style – she has managed to alert the global community to the Anthropocene, its essence and its implications. Thoreau would have been amused, and so would a distant relative of Greta, Svante Arrhenius, the Nobel Prize-winning physicist who pioneered studies of the relationship between carbon dioxide in the atmosphere and global temperature.

Opposite above: Extinction Rebellion in Trafalgar Square, London.

Opposite below: Tokyo kindergarten designed by Takaharu Tezuka and Yui Tezuka

Below: Greta Thunberg launching the first climate strike at Swedish Parliament, Stockholm, 2018.

Overleaf: School strike in Sydney, Australia, 15 March, 2019.

TWENTY-SEVEN

....

Geopolitics as housekeeping

Developed in the sixteenth century, the term "housekeeping" denoted activities for maintaining a household, emphasizing the necessity of making ends meet in a very broad sense, environmentally and financially. Several related terms – some derived from the ancient Greek notion of *oikos*, the "household" of life – draw attention to the maintenance of economies and ecosystems. Collectively, the Greek households made up the *polis*, the governing unit of the city and the state. No doubt, many non-Western traditions have similar notions of household and community, represented in the East Asian context, for example, by the Chinese terms *jia*, *guo* and *tianxia*.[1]

In the ancient world, then, politics was a fairly broad phenomenon, both small-scale and large-scale (household and state) pertaining at the same time to the natural as well as the social domain (the land and its resources as well as the lives of the family and the citizens of the state). Unfortunately, in modern environmental language, *oikos* is often stripped of its conflation with social life. Thus, AB Hollingshead, whose ideas helped to develop human ecology, spoke in the 1940s of "the ecological and social orders": "The former is primarily an extension of the order found everywhere in nature, whereas the latter is exclusively ... a distinctly human phenomenon".[2]

In such a scheme, the term "geopolitics" has usually been narrowly restricted to regional or international politics – to power and governance as they pertain to geographic zones and continents. It now seems pertinent, given the all-encompassing Anthropocene, to advance a broader notion of geopolitics, which includes not just humans but the Earth itself.[3] Such an approach to politics involves attending to the planet in the broadest sense – including volcanoes, rivers, minerals, glaciers, etc. – as well as to the social life that unfolds on the surface. An active volcano, as we have seen, is itself quite lively, sometimes as a result of human activities. Clearly, in the Anthropocene, geological activities, eruptions and earthquakes are matters of the new geopolitics.

Given the nature and scale of ongoing environmental change, then, it seems pertinent

Opposite: *The Blue Marble* image of Earth captured by the crew of Apollo 17 spacecraft on 7 December, 1972.

GEOPOLITICS AS HOUSEKEEPING

to revert to the classic notion of housekeeping, expanding its reference.[4] In the (gendered) language of American palaeontologist Henry Fairfield Osborn (*Our Plundered Planet*, 1948), "it is man's earth now". This statement is even more relevant now than in Osborn's time. A politics of collective responsibility and stewardship, if not "ownership", needs to replace the fragmenting privatization and commodification of recent decades of the Earth and its resources, which have driven the house into the "on fire" mode. Without collective concerns and action, the house remains permanently on alert. Significantly, a recent influential Anthropocene manifesto – the work of the Swedish Thunberg family, informed by school strikes and Extinction Rebellion – is entitled *Our House is on Fire: Scenes of a Family and a Planet in Crisis*.[5] The planet and the house firmly unite on the global environmental and governmental agenda.

No doubt the new perspective of global housekeeping has been partly triggered by the view from space, especially *The Blue Marble* image of Earth (no. AS17-148-22727 in the records of NASA) captured by the crew of the Apollo 17 spacecraft on its way to the Moon on 7 December 1972. More importantly, though, recent environmental crises associated with superheating, freakish weather, floods, and firestorms have pushed for a change in perspective. What are the implications of such a change in geopolitics?

Geopolitics in the new sense is now being fleshed out for the Anthropocene on many fronts. It's a difficult task, keeping in mind the novelty of the situation, the planetary scales, different interests (north and south and all kinds of social divisions), the pressure of time, frightening levels of CO_2 in the atmosphere, the alarming role of denial and fake news, and the complications of seeking global agreements on key issues. While human knowledge of both the natural world and the possibilities for action have been vastly expanded, one often feels hopeless confronting the gigantic tasks involved. Obviously, a broad collaboration is needed, not only of different academic disciplines but also the arts, and across all kinds of levels (geographic, political, and environmental).

It may be time to bring to the agenda the notion of solidarity, which has a long history in social thought and politics. Austrian political scientist Barbara Prainsack has usefully outlined its history and growing salience in recent years.[6] It seems that solidarity – the communitarian spirit that negates purely private or self-centred interests to facilitate necessary collective action for planetary life – is of absolutely vital geopolitical importance during the Anthropocene. Etymologically, the notion of "solidarity" derives from the concept of *in solidum* in Roman law, which referred to a joint contract. The concept of "solid" – as in "solid rock" – has the same roots, stemming from *solidus*, meaning "firm, whole, undivided, entire". Interestingly, a recent trend in architecture, launched by architect Amin Taha, stonemason Pierre Bidaud, and engineer Steve Webb, is described as "The New Stone Age", characterized by reintroducing stone, the forgotten building material of our time, for the purpose of reducing construction costs and for minimizing the carbon footprint by avoiding steel and concrete.[7]

Indeed, the house and the home is a good place to start rethinking geopolitics. Extending the horizon, feminist scholars who use the shared pen name J.K. Gibson-Graham have argued that what is needed in the Anthropocene is a new ethics of care, relating to the global world as one does to a family or a house. Can we extend our solidarity, they ask, to the more-than-human, to other life forms and life in general? "If we can", they conclude, "that would certainly usher in a new ... form of belonging".[8]

Anthropocenic solidarity, by definition, includes the planet and life itself. A liveable future is all about planetary relations. Human activity, however, must be considered a key "driver" of global environmental change. The impact of the human footprint on Earth has begun to equal that of (other) geological forces. Herodotus (writing in the fifth century BCE) remarked in his *Histories*: "Of all man's miseries, the bitterest is this, to know so much and to be impotent to act". This is no less relevant now than in Herodotus's time. Concerted action is crucial. Once again, the cover image of *The Economist* from 2011 comes to mind, with its "Welcome to the Anthropocene" banner. This is the predicament of modern housekeeping, on a new scale. Concrete proposals are now being crafted for the geopolitical agenda, under immense pressure, including the Green New Deal outlined by Naomi Klein, Ann Pettifor, and some others in the US, Canada, UK and elsewhere, emphasizing structural changes in finance and the economy, informed by notions of solidarity, equity and well being.[9]

Opposite: An oil tanker, with an oil rig in the distance.

Studies of tipping points – the points of no return – are essential for current geopolitics, for evaluating the state of the planet and the scale of the environmental emergency. In particular, it is important to establish the level of reduction of CO_2 emissions needed to avoid the melting of glaciers, sea-level rises and weather extremes, and the appropriate avenues for geopolitical action. The conclusion of the scientific community, summarized in a commentary in *Nature* in 2019, is unequivocal. While previously, scholars assumed there was a low probability of reaching tipping points, such as the loss of the Amazon rainforest, evidence is mounting that such events "could be more likely than was thought, have high impacts and are interconnected across different biophysical systems, potentially committing the world to long-term irreversible changes".[10] Two decades ago, the Intergovernmental Panel on Climate Change (IPCC) indicated that "large-scale discontinuities" in the climate system were to be expected only if global warming exceeded 5 degrees Celsius above pre-industrial levels, but the most recent IPCC Special Reports (published in 2018 and 2019) suggest that tipping points could be exceeded even in the case of warming between 1 and 2 degrees Celsius.

This is an alarming conclusion, considering the window of opportunities at our disposal. One can envisage the global household – the Earth's System, in ecological jargon – as a super-tanker at sea, perhaps an oil tanker with dangerous cargo, desperately trying to shift to a safe course out of respect for a heating planet. Will there be time for the tanker, given its titanic momentum, to turn or to halt before it is too late?

One of the authors of the *Nature* commentary of 2019, Owen Gaffney, a global sustainability analyst at the Stockholm Resilience Centre at Stockholm University, offered this view: "I don't think people realize how little time we have left … We'll reach 1.5° C in one or two decades, and with three decades to decarbonise it's clearly an emergency situation".[11]

Unfortunately, the 2019 UN talks in Madrid did not provide much space for optimism, ending in geopolitical deadlock. How can humanity deal with the colossal and critical task of shifting the course of the tanker, including changing the direction of public discourse, the academic community and the planet itself – before reaching or passing the tipping points? What would be good governance in such a context? How is it possible to align public and private interests, the material and the organic, the local and the global, governments and the international arena? Such questions – some of the most pressing questions ever faced by humanity – are likely to remain central concerns. New governance models are needed for dealing with the crisis, aligning a variety of actors and agencies at different levels of governance under conditions of complexity and uncertainty.

The Geopolitical Assembly of the Earth, if such a thing can be imagined, consists of the planet, representatives of diverse groups of people from all over the world, and organisms of all shapes and sizes. Mass protests and general strikes may break out at any time, like swarms, transcending all borders. Events are difficult to understand or manage, as we see in the case of international climate-change conferences, but this is the world which people, other living beings and the planet have to live with, and it is as well to adjust to it.[12] Humanity is part of the nature that we describe and grapple with; we are bound up together in place and time, like the seismometers that sense earthquakes and predict eruptions, in constant connection with the Earth and dancing with its movements.

Opposite: Glacier as seen from a NASA satellite.

Above: Extinction Rebellion's *Sinking House*, a protest art-installation in the Thames. 17 November, 2019.

Following pages: A man comforts his dog outside a family home devastated by bush fires, South-West Victoria, Australia, 2019.

....

A fuming planet

The Triumph of Death, an oil painting by Pieter Bruegel the Elder, c.1562, is often considered one of the most terrifying works in the history of art. It depicts an existential threat to humanity, an advancing plague with near-total mortality. Bruegel's canvas is littered with skeletons, agonized bodies, battling armies, public gatherings and the tools of execution. The land is dry, and fires are burning near and far. The bell in the top left corner signals the end of the world. In the bottom corner, to the right, a troubadour and a singer entertain each other, both oblivious to the grim reality around them. It seems the social order is collapsing. It is tempting to see Bruegel's world as representing the current Anthropocene, the panic surrounding it and its depressing future. Fears of crisis, as Russian literary scholar Mikhail Bakhtin pointed out, were part of ancient mythologies where "cosmic terror" reigned: "the fear of the immeasurable, the infinitely powerful".[1] It is tempting to think of the "Welcome to the Anthropocene" imagery discussed at the beginning of this book. In any case, our planet is fuming.

It seems vitally important to come to terms with the "cosmic terror" of the Anthropocene, which is infinitely more powerful than anything else in human history – certainly more powerful than a single plague involving a single species. In the Middle Ages, Bakhtin suggested, people would often respond to terror with laughter – that would hardly be the best strategy now, although it might help alleviate stress. The assumption that the history of Earth has now entered a new epoch, the Anthropocene, is a useful beginning. The next step is to establish the nature and scale of the crisis, to raise alarms and to move on.

It is important to keep in mind that, as we go back through history, our data sets (based on historical records, ice cores and tree rings) become increasingly difficult to compare to recent data, so that any extrapolation into the future is riddled with uncertainty. However, the available evidence is highly suggestive, and discounting it would be extremely risky for life on the planet.[2] In order to deal with the risks and uncertainties involved, as well as deciding on the relevant courses of action, all kinds of sciences, arts and agencies are needed. Critical interrogation of central concepts in current environmental debates is essential; without it, there would be no way of knowing whether or not we are on the "right" track. We will never get completely out of the metaphoric traps of language, but some languages are better than others.

The policy of "winner-takes-all", silencing minorities and the marginalized, is obviously not constructive, for either environmental science or politics. The history of science demonstrates that scientific knowledge only advances as long as there is space for honest critical debates. Otherwise, the Earth would still be considered flat and everything celestial would rotate around it. On the other hand, it is risky to abandon the consensus established by the scientific community, after thorough research and intensive discussions. The public has to be able to trust the judgement of the academic community and its procedures for establishing planetary "truths", however flawed they may be, independent of vested interests such as those of the carbon industry, established aristocracy, religious orthodoxy, the patriarchy and white supremacy.

Scientists have recently accepted culpability for tending "to underestimate the severity of threats and the rapidity with which they might unfold", one reason being "the perceived need for consensus".[3] Making a deal with the climate deniers and the detractors, for the sake of finding an imaginary middle ground and for keeping

Below: *The Triumph of Death* (c.1562), by Pieter Bruegel the Elder.

the peace, carries extreme risks for life on the planet. The scale of the current environmental crisis demands new kinds of social institutions and collaboration, robust and flexible enough to generate the necessary trust of and cooperation between academic disciplines, interest groups, national governments and international bodies.

Historical case studies of climate change can be as revealing as massive data sets about greenhouse gases and global temperatures. Interestingly, one of the most instructive case studies involves not heating but cooling – the so-called Little Ice Age, a somewhat misleading term – and human responses to it. In the thirteenth century, parts of the Northern Hemisphere began cooling for several reasons, including volcanic eruptions and minute decreases in solar radiation. Temperatures were suppressed until persistent warming set in during the nineteenth century. While some societies were unprepared, suffering losses, if not collapses, others adapted and pulled through. Thus, resilient Dutch society, with its diverse diet, commercial fleets and urban charities, was flexible enough to transform itself, booming for much of the sixteenth and seventeenth centuries – a period which is remembered as the Dutch "Golden Age".[4]

Environmental historian Dagomar Degroot argues that cases such as this urge us to "approach the future with open minds ... to implement radical policies ... that go beyond simply preserving what we have now, and instead promise a genuinely better world for our children".[5] The optimistic spirit that Degroot implies is highlighted in Hendrick Avercamp's painting *Winter Landscape with Ice Skaters* (c.1608), which captures a Dutch community literally dancing on ice, playfully interacting on the streets despite the chilly

climate and the challenges it entailed. What a striking contrast to Bruegel's depressing take in *The Triumph of Death* half a century earlier!

The decisive challenge of the politics of life in the Anthropocene is to avert total disaster and

hope for the best. Meaningful geopolitics for the future – in the new, dual sense of the global and the geologic – needs a deep understanding of endangered lives and environments, the forms that solidarity might take, careful planning around what to prioritize and how to act, and, of course, endless optimism, playfulness and diplomacy.

Above: *Winter Landscape with Ice Skaters* (c.1608), by Hendrick Avercamp.

T I M E L I N E
....

Before the Common Era (BCE) and after the Common Era (CE)

200,000 BCE
The origin of the use of fire

9500
First cultivation of major crops

6300
Short-lived global cooling

2250
Massive global droughts

1492 CE
Columbian exchange begins

1500
Plantation slavery

1712
The development of the first
steam engine, in England

1778
Publication of the Comte de Buffon's
The Epochs of Nature.

1800
The Industrial Revolution

1820s
Jean-Baptiste Joseph Fourier hypothesizes the
effects of the Sun on the Earth's atmosphere,
foreshadowing the "greenhouse effect"

1858
Darwin and Wallace announce
their theory of evolution

1860s
The birth of "extinction"

1865
The beginning of fracking

1896
Svante Arrhenius describes greenhouse effect
and predicts increase in CO_2 and global heating

1900
The development of plastic

1913
Charles Fabry and Henri Buisson
discover the ozone layer

1936
Alan Turing launches computer science

1938
Guy Stewart Callendar demonstrates rise in CO_2

1945
First atomic bomb explosion, New Mexico

1952
Rosalind Franklin photographs DNA

1954

Early production of solar panels

1962

Rachel Carson's book *Silent Spring* is published

1967

Syukuro Manabe and Richard T. Wetherald create the first computer model to simulate the climate of Earth

1969

Humans land on the Moon

1971

Signing of the Ramsar Convention on Wetlands

1972

The Blue Marble image of Earth, 7 December

1979

Apollo 11 landing on the Moon

1980s

Arrival of Anglo-American neoliberalism

1981

James E. Hansen and colleagues establish human-caused global warming

1986

Chernobyl disaster

1988

Intergovernmental Panel on Climate Change (IPCC) established

1992

The first Earth Summit, Rio de Janeiro

1997

Captain Charles Moore discovers plastic soup in the Pacific Ocean

2000

The proposal of the term "Anthropocene"

2006

Plastiglomerate, a fusion of plastic and natural sediment, is discovered on Kamilo Beach, Hawaii

2016

The birth of "fake news"

2018

Founding of Extinction Rebellion; Greta Thunberg sparks youth protests worldwide

2019

The United Nations issue a report on extinction, titled "One million threatened species?"

2019

Devastating bush fires and record air temperatures in Australia; funerals held for dead glaciers in Iceland and Switzerland

2020

Covid-19 crisis signals huge reduction in air traffic and carbon emissions; it remains unknown at the time of writing whether these effects will be ongoing

E N D N O T E S

....

1 Introduction: A new epoch

1 Dalby, Simon. 2016. "Framing the Anthropocene: The good, the bad and the ugly". *The Anthropocene Review* 3(1): 33–51. DOI: 10.1177/2053019615618681.
2 Sklair, Leslie. 2019. "Globalization and the challenge of the Anthropocene". In *Globalization*, edited by Ino Rossi. New York: Springer.
3 Sklair, Leslie. 2018. "The Anthropocene Media Project: Mass media human impacts of the Earth System". *Visions for Sustainability* 10. DOI: 10.13135/2384-8677/2740.
4 Quammen, David. 2020. "We made the Coronavirus epidemic". *New York Times*, 28 January, 2020.
   Vidal, John. 2020. "'Tip of the iceberg': Is our destruction of nature responsible for Covid-19?" *The Guardian*, March 18, 2020.
5 Sanson, Ann. 2020. "Venice canals appear cleaner amid coronavirus lockdown". *The Art Newspaper*, March 17, 2020.

2 Challenges to the Anthropocene

1 Frisch, Max. 1980. *Man in the Holocene: A Story*. London: Dalkey Archive Press.
2 Worster, Donald ed. 1988. *The Ends of the Earth: Perspectives on Environmental History*. Cambridge: Cambridge University Press.
3 Marsh, George Perkins. (1864) 1965. *Man and Nature, or Physical Geography*. Cambridge: The Belknap Press of Harvard University Press.
4 Steffen, Will, Jacques Grinevald, Paul Crutzen and John McNeill. 2011. "The Anthropocene: Conceptual and historical perspectives". *Philosophical Transactions of the Royal Society* 369: 842–867.
5 Leclerc, Georges-Louis. 2018. *The Epochs of Nature*. Translated by Jan Zalasiewicz, Anne-Sophie Milon, Mateusz Zalasiewicz. Chicago: University of Chicago Press
   Glacken, Clarence J. 1967. *Traces on the Rhodian Shore: Nature and Culture in Western Thought from Ancient Times to the End of the Eighteenth Century*. Berkeley: University of California Press.
6 Pálsson, Gísli, Szerszynski, Bronislaw, Sörlin, Sverker et al. 2013. "Reconceptualizing the 'Anthropos' in the Anthropocene: Integrating the social sciences and humanities in global environmental change research". *Environmental Science and Policy* 28: 3-13.
7 Kolbert, Elizabeth. 2019. "Age of Man: Enter the Anthropocene". *National Geographic*, July 5, 2019.
8 Zalasiewicz, Jan, Williams, M., Haywood, A., Ellis, M., 2011. "The Anthropocene: A new epoch of geological time?" *Philosophical Transactions of the Royal Society A: Mathematical, Physical and Engineering Sciences* 369: 835-841.
9 Showstack, Randy. 2013. "Scientists debate whether the Anthropocene should be a new geological epoch". *Eos* 94(4): 41-42.
10 Meyer, Robinson. 2018. "Geologic timekeepers are feuding: 'It's a bit like Monty Python." *The Atlantic*, July 20, 2018.
11 Brannen, Peter. 2019. "The Anthropocene is a joke". *The Atlantic*, August 14, 2019.
   Santana, Carlos. 2019. "Waiting for the Anthropocene". *British Journal for the Philosophy of Science* 70: 1073–1096.

3 The recognition of deep time

1 McPhee, John. 1981. *Basin and Range*. New York: FSG
   Rudwick, Martin J.S. 2014. *Earth's Deep History: How It Was Discovered and Why It Matters*. Chicago: The University of Chicago Press.
   Macfarlane, Robert. 2019. *Underland: A Deep Time Journey*. London: W.W. Norton & Company.

2 Winchester, Simon. 2001. *The Map That Changed the World*. London: Penguin Books.
3 Shubin, Niels. 2019. "Extinction in deep time: Lessons from the past". In *Biological Extinction: New Perspectives*, edited by Partha Dasgupta, Peter H. Raven and Anna L. McIvor, 22–33. Cambridge: Cambridge University Press.
4 Torrens, Hugh. 1995. "Mary Anning (1799–1847) of Lyme: The Greatest Fossilist the World Ever Knew". *The British Journal for the History of Science* 25(3): 257–284.
5 Hutton, James. 1788. "X. Theory of the Earth; or an Investigation of the Laws Observable in the Composition, Dissolution and Restoration of Land upon the Globe". *Transactions of the Royal Society of Edinburgh* 1 (2). Royal Society of Edinburgh Scotland Foundation: 209–304. doi:10.1017/S0080456800029227.
6 Wulf, Andrea. 2015. *The Invention of Nature: The Adventures of Alexander von Humboldt*. London: John Murray.
7 Gorman, James. 2019. "Humans dominated Earth earlier than previously thought". *New York Times*. 3 September, 2019.

4 Early signs and warnings

1 Wikipedia. n.d. "History of climate change science". Accessed 13 November 2019. https://en.wikipedia.org/wiki/History_of_climate_change_science.
2 Mayewski, Paul Andrew and Frank White. 2002. *The Ice Chronicles: The Quest to Understand Global Climate Change*. Hanover: The University Press of New England.
3 Hansen, J., et al. 1981. "Climate impact of increasing atmospheric carbon dioxide". *Science* 213: 957-966 doi:10.1126/science.213.4511.957.
4 Lovelock, James E. 1979. *Gaia: A New Look at Life on Earth*. Oxford: Oxford University Press.

5 Fire and the Long Anthropocene

1 Eiseley, Loren. 1978. "Man the Firemaker". In *The Star Thrower*. New York: Hartcourt Brace Jovanovich.
2 Longrich, Nick. 2019. "Were other humans the first victims of the sixth mass extinction?" *The Conversation*, November 21, 2019.
3 Gorman, James. 2019. "Humans dominated Earth earlier than previously thought". *New York Times*, September 3, 2019.
4 Pyne, Stephen J. 2015. "The Fire Age". *Aeon*, May 5, 2015.
5 Crosby, Alfred. 1986. *Ecological Imperialism: The Biological Expansion of Europe, 900–1900*. Cambridge: Cambridge University Press.
6 Pálsson, Gísli. 2016. *The Man Who Stole Himself: The Slave Odyssey of Hans Jonathan*. Translated by Anna Yates. Chicago: University of Chicago Press.

6 The sad fate of disappearing species

1 Braun, Ingmar M. 2018. "Representations of birds in the Eurasian Upper Palaeolithic Ice Age Art". *Boletim do Centro Português de Geo-História e Pré-História* 1(2): 13–21.
2 Grieve, Symington. (1885) 2015. *The Great Auk, or Garefowl*. Cambridge: Cambridge University Press.
3 Pálsson, Gísli. 2020. *Fuglinn sem gat ekki flogið*. Reykjavík: Mál og menning.
4 Wolley, John. 1858. *Gare-Fowl Books*. Manuscript. From Cambridge University Library. MS Add. 9839/2/1–5/1. hl., 110.
5 Kingsley, Charles. (1863) 2003. *The Water-Babies*. Nottinghamshire: Award Publications Ltd.
6 Thomas, Jessica E. et al. 2019. "Demographic reconstruction from ancient DNA supports rapid extinction of the great auk". *eLife* 2019;8:e47509. DOI: 10.7554/eLife.47509.

7   The birth of extinction and endlings

1   Greene, John. 1959. *The Death of Adam: Evolution and Its Impact on Western Thought*. Ames: Iowa State University.

2   Leclerc, Georges-Louis. 2018. *The Epochs of Nature*. Translated by Jan Zalasiewicz, Anne-Sophie Milon, Mateusz Zalasiewicz. Chicago: University of Chicago Press

3   Newton, Alfred. 1896. "Extermination". In *Dictionary of Birds*, 214–229. London: Adam and Charles Black.

4   Cowles, Henry M. 2013. "A Victorian extinction: Alfred Newton and the evolution of animal protection". *British Society for the History of Science* 46(4): 695–714.

5   Bargheer, Stefan. 2018. "The sociology of morality as ecology of mind: Justifications for conservation and the international law for the protection of birds in Europe". *European Journal of Sociology* 59(1): 63–89.

6   Wollaston, A.F.R. 1921. *Life of Alfred Newton*. London: John Murray.

7   Marsh, George Perkins. 1867. *Man and Nature: On Physical Geography as Modified By Human Action*. New York: Charles Schribner & Co.

8   Jørgensen, Dolly. 2017. "Endling, the power of the last in an extinction-prone world". *Environmental Philosophy* 14(1): 119–138. doi:10.5840/envirophil201612542
     Nijhuis, Michelle. 2017. "What do you call the last of a species?" *The New Yorker*, March 2, 2017.

9   Jacobs, Julia. 2019. "George the snail, believed to be the last of his species, dies at 14 in Hawaii". *New York Times*, January 10, 2019.

8   Enter the Industrial Revolution

1   Simmons, I.G. 1989. *Changing the Face of the Earth: Culture, Environment, History*. Oxford: Basli Blackwell.

2   Dickens, Charles. 1850. *David Copperfield*. London: Bradbury & Evans.

3   Merchant, Caroline. 2006. "The Scientific Revolution and The Death of Nature". *Isis* 97(3): 513–533.

4   Schwab, Klaus. 2015. "The Fourth Industrial Revolution". *Foreign Affairs*, December 12, 2015.

5   Wired. 2015. "8 cities that show you what the future will look like". Accessed December 19, 2019. https://www.wired.com/2015/09/design-issue-future-of-cities/

9   The atomic age

1   Battaglia, Debbora, ed. 2005. *E.T. Culture: Anthropology in Outerspaces*. Durham: Duke University Press
     Finney, Ben R. and Eric M. Jones, eds. 1985. *Interstellar Migration and the Human Experience*. Berkeley: University of California Press.

2   Than, Ker. 2016. "The Age of Humans: Living in the Anthropocene". *Smithsonian Magazine*, January 7, 2016.

3   Fiorini, Ettore. 2014. "Nuclear energy and Anthropocene". *Rendiconti Licncei: Scienze Fisiche e Naturali* 25: 119–126.

4   Simmons, I.G. 1989. *Changing the Face of the Earth: Culture, Environment, History*. Oxford: Basil Blackwell.

5   Alexievich, Svetlana. 2005. *Voices from Chernobyl: The Oral History of a Nuclear Disaster*. Translated by Keith Gessen. Champaign: Dalkey Archive Press.

6   Petryna, Adriana. 2013. *Life Exposed: Biological Citizens after Chernobyl*. Princeton: Princeton University Press.

7   Ibid.

8   Codignola, Luca and Schrogl, Kai-Uwe eds. 2009. *Humans in Outer Space: Interdisciplinary Odysseys*. New York: Springer.

10   Draining wetlands

1   Huijbens, Edward and Gísli Pálsson. 2009. "The bog in our brain and bowels: Social attitudes to the cartography of Icelandic wetlands". *Environment and Planning D: Society and Space* 27: 296-316
     Fraser, L.H. and P.A. Keddy. 2005. "The future of large wetlands: A global perspective". In *The World's Largest Wetlands: Ecology and Conservation*, edited by L.H. Fraser and P.A. Keddy, 446-68. Cambridge: Cambridge University Press.
     Mundy, Vincent. 2019. "Mother Nature recovers amazingly fast: Reviving Ukraine's wetlands". *The Guardian*, December 27, 2019.

2   Swift, Graham. 1983. *Waterland*. New York: Washington Square Press.

3   Thoreau. 1856. Quoted in H. Prince. 1997. *Wetlands of the American Midwest: A Historical Geography of Changing Attitudes*. Chicago: The University of Chicago Press.

4   Constanza Robert et al. 1997. "The value of the world's ecosystem services and natural capital". *Nature* 387: 253-260.

5   White, R. 1996. *The organic machine*. New York: Hill and Wang.

6   Mitsch, W.J. and J.G. Gosselink. 2007. *Wetlands*. Hoboken: John Wiley & Sons. P. 353.

7   Strang, Veronica. 2005. "Common senses: Water, sensory experience and the generation of meaning". *Journal of Material Culture* 10: 92-120.

8   Huijbens, Edward H. and Gísli Pálsson. 2009. "The bog in our brain and bowels: Social attitudes to the cartography of Icelandic wetlands". *Environment and Planning D: Society and Space* 27: 296-316.

9   Mundy, Vincent. 2019. "Mother Nature recovers amazingly fast: Reviving Ukraine's wetlands". *The Guardian*, December 27, 2019.

10   IPCC. 2019. *Climate Change and Land*. Accessed December 9, 2019. https://www.ipcc.ch/site/assets/uploads/2019/08/4.-SPM_Approved_Microsite_FINAL.pdf.

11   Plastics: Broth, soup and islands

1   Meloni, Maurizio. 2019. *Impressionable Biologies: From the Archaeology to Plasiticity to the Sociology of Epigenetics*. New York: Routledge.

2   Abbing, Michiel Roscam. 2019. *Plastic Soup: An Atlas of Ocean Pollution*. Washington: Island Press.

3   Schlanger, Zoë. 2019. "Yes, there's microplastic in the snow". *Quartz*. Accessed December 25, 2019. https://www.sciencedirect.com/science/article/pii/S221330541930044X?via percent3Dihub.

4   Carrington, Damian. 2019. "Revealed: Microplastic pollution raining down on city dwellers". *The Guardian*, December 27, 2019.

5   Carrington, Damian. 2019. "After bronze and iron, welcome to the plastic age, say scientists". *The Guardian*, September 4, 2019.

6   Smith, Roberta, n.d. Quoted in Christie's. 2019. "Rabbit by Jeff Koons – a chance to own the controversy". *Christie's*. Accessed December 28, 2019. https://www.christies.com/features/Jeff-Koons-Rabbit-Own-the-controversy-9804-3.aspx. Opened on

7   Created with a team of local volunteers and artisans including Cyntault Creations, Tamsen Rae, Clara Cloutier and Jean-Michel Cholett.

8   Wikipedia. n.d. "Plastiglomerate". Accessed December 27, 2019. https://en.wikipedia.org/wiki/Plastiglomerate.

12   Superheating

1   Worster, Donald ed. 1988. *The Ends of the Earth: Perspectives on Environmental History*. Cambridge: Cambridge University Press.

2   Leclerc, Georges-Louis. 2018. *The Epochs of Nature*. Translated by Jan Zalasiewicz, Anne-Sophie Milon, Mateusz Zalasiewicz. Chicago: University of Chicago Press.

3   Leclerc, Georges-Louis. 2018. *The Epochs of Nature*. Translated by Jan Zalasiewicz, Anne-Sophie Milon, Mateusz Zalasiewicz. Chicago: University of Chicago Press.

4   Zalasiewicz, Jan et al. 2018. "Introduction". In *The Epochs of Nature*. Translated and edited by Jan Zalasiewicz, Anne-Sophie Milon and Mateusz Zalasiewicz, p.xiv. Chicago: University of Chicago Press.

5   Fountain, Henry and Nadja Popovich. 2020. "2019 was the second-hottest year ever, closing out the warmest decade". *New York Times*, January 15, 2020.

6   Nuccitelli, Dana. 2019. "Climate models have accurately predicted global heating, study finds". *The Guardian*, December 4, 2019.

7   Fagan, Brian. 2008. *The Great Warming: Climate Change and the Rise and Fall of Civilizations*. New York: Bloomsbury Press.

8   Eriksen, Thomas Hylland. 2016. *Overheating: An Anthropology of Accelerated Change*. London: Pluto Press.

9   Karlsruhe Institute of Technology. 2019. "Crowd oil – fuels from air-conditioning". *Phys.Org*. Accessed May 6 2019. https://phys.org/news/2019-05-crowd-oilfuels-air-conditioning.html.

13   The end of glaciers

1   Ladurie, Emmanuel Le Roy. 1971. *Times of Feast, Times of Famine: A History of Climate Since the Year 1000*. New York: Doubleday.
     Cruikshank, Julie. 2005. *Do Glaciers Listen?: Local Knowledge, Global Encounters, & Social Imagination*. Vancouver: UBC Press.

2   Orlove, Benjamin, Ellen Wiegandt and Brian H. Luckman eds. 2008. *Darkening Peaks: Glacier Retreat, Science, and Society*. Berkeley: University of California Press.

3   Galey, Patrick. 2019. "Amazon fires 'quicken Andean glacier melt.'" *Phys.org*. Accessed November 28, 2019. https://phys.org/news/2019-11-amazon-andean-glacier.html

4   Adhikari, Surendra and Erik R. Ivins. 2016. "Climate-driven polar motion: 2003 – 2015". *Science Advances* 8 April, 2: e1501693.
5   Goodell, Jeff. 2019. "Why Venice is disappearing". *Rolling Stone*. Accessed: November 15, 2019. https://www.rollingstone.com/politics/politics-news/venice-flooding-2019-mose-corruption-913175/
6   McDougall, Dan. 2019. "'Ecological grief: Greenland residents traumatized by climate emergency". *The Guardian*, August 12, 2019.
7   Taylor, Alan. 2016. "Peru's Snow Star Festival". *The Atlantic* June 7, 2016 Wikipedia. n.d. *Quyllurit'i*. Accessed January 13, 2020. https://en.wikipedia.org/wiki/Quyllurit%27i.
8   Siad, Arnaud and Amy Woodyatt. 2019. "Hundreds mourn 'dead' glacier at funeral in Switzerland". CNN, September 22, 2019. Accessed January 13, 2020. https://edition.cnn.com/2019/09/22/europe/swiss-glacier-funeral-intl-scli/index.html.
9   Björnsson, Helgi. 2017. *The Glaciers of Iceland: A Historical, Cultural and Scientific Overview*. New York: Springer.
10   Mayewski, Paul Andrew and Frank White. 2002. *The Ice Chronicles: The Quest to Understand Global Climatic Change*. Hanover, NH: University Press of New England.
11   World Ocean Observatory. 2019. "A stunning art installation showing projected sea-level rise". *World Ocean Forum*. Accessed: March 13, 2019. https://medium.com/world-ocean-forum/a-stunning-art-installation-showing-projected-sea-level-rise-fc05ef1825cd.

14   Freakish weather
1   Strauss, Sarah and Benjamin S. Orlove eds. 2003. *Weather, Climate, Culture*. Oxford: Berg Publishers.
2   Foer, Jonathan Safran. 2019. *We Are the Weather: Saving the Planet Begins at Breakfast*. New York: Farrar, Straus and Giroux.
3   Ogilvie, Astrid E.J. and Gísli Pálsson. 2003. "Mood, magic and metaphor: Allusions to weather and climate in the *Sagas of Icelanders*". In *Weather, Climate, Culture*, edited by Sara Strauss and Benjamin S. Orlove, 251–274. Oxford: Berg Publishers.
4   Wikipedia. n.d. "Tropical cyclone". Accessed: January 22, 2020. https://en.wikipedia.org/wiki/Tropical_cyclone.
5   Mir Emad Mousavi, Jennifer L. Irish, Ashley E. Frey et al. 2011. "Global warming and hurricanes: The potential impact of hurricane intensification and sea level rise on coastal flooding". *Climatic Change* 104: 575–597.
6   Nevárez, Julia. 2018. *Governing Disaster in Urban Environments: Climate Change Preparation and Adaption after Hurricane Sandy*. New York: Lexington Books Wikipedia. n.d. "Hurricane Katrina". Accessed January 22, 2020. https://en.wikipedia.org/wiki/Hurricane_Katrina
7   Masters, Jeff. 2019. "A Review of the Atlantic Hurricane Season of 2019". *Scientific American*, November 25, 2019.
8   Flavelle, Christopher. 2020. "Conservative states seem billions to brave for disaster. Just don't call it climate change". *New York Times*, January 20, 2020.
9   Masco, Joseph. 2010. "Bad weather: On planetary crisis". *Social Studies of Science* 40(1): 7–40. P. 9.
10   Ibid. P. 26.

15   Volcanic eruptions
1   Ellsworth, William L. 2013. "Injection-induced earthquakes". *Science* 341 (6142). DOI: 10.1126/science.1225942.
2   Pálsson, Gísli. 2020. *Down to Earth: A Memoir*. Galeto: Punctum Books.
3   Sneed, Annie. 2017. "Get ready for more volcanic eruptions as the planet warms". *Scientific American*, December 21, 2017 Swindles, Graeme T., Elizabeth J. Watson, Ivan P. Savov et al. 2018. "Climate control on Icelandic volcanic activity during the mid-Holocene". *Geology* 46(1): 47–50.
4   Pagli, Carolina and Freysteinn Sigmundsson. 2008. "Will present day glacier retreat increase volcanic activity? Stress induced by recent glacier retreat and its effect on magmatism at the Vatnajökull Ice Cap, Iceland". *Geophys. Res. Lett.*, 35: 1–5.
5   Compton, Kathleen, Richard A. Bennett and Sigrún Hreinsdóttir. 2015. "Climate-driven vertical acceleration of Icelandic crust measured by continuous GPS geodesy". *Geophysical Research Letters* 42(3), 743–750. https://doi.org/10.1002/2014GL062446 Goldenberg, Suzanne. 2015. "Climate change is lifting Iceland". *The Guardian*, January 30, 2015 Anon. 2015. "Iceland rises as its glaciers melt from climate action".

Astrobiology Magazine. January 30, 2020.
6   Holmberg, Karen. 2020. "Inside the Anthropocene volcano". In *Critical Zones: The Science and Politics of Landing on Earth*, edited by Bruno Latour and Peter Weibel. Cambridge: MIT Press. Halperin, Ilana, et al. "Hand held lava: On the sight of towering, pyramid rocks, from Hawaii to Vesuvius to Grimsvotn". Performance art hosted by Triple Canopy, Brooklyn, New York, October 8, 2010, https://www.canopycanopycanopy.com/contents/hand_held_lava.
7   Hayoun, Nelly Ben. 2010. *The Other Volcano*. Accessed January 16, 2020. http://nellyben.com/projects/the-other-volcano/

16   The fragile ocean
1   Probyn, Elsbeth. 2016. *Eating the Ocean*. Durham: Duke University Press.
2   Fagan, Brian. 2012. *Beyond the Blue Horizon: How the Earliest Mariners Unlocked the Secrets of the Oceans*. New York: Bloomsbury Press.
3   Astrup, Poul, Peter Bie and Hans Chr. Engell. 1993. *Salt and Water in Culture and Medicine*. Copenhagen: Munksgaard.
4   Stott, Rebecca. 2000. "Through a glass darkly: Aquarium colonies and nineteenth-century narratives of marine monstrosity". *Gothic Studies* 2(23): 305–327.
5   Pálsson, Gísli. 2006. "Nature and society in the age of postmodernity". In *Reimagining Political Ecology*, edited by Aletta Biersack and James Greenberg, 70–93. Durham: Duke University Press.
6   McCay, Bonnie J. and James M. Acheson eds. 1987. *The Question of the Commons: The Culture and Ecology of Communal Resources*. Tucson: University of Arizona Press.
7   Carson, Rachel. (1955) 1998. Illustrations Robert W. Hines. Boston: Mariner Books. Lepore, Jill. 2018. "The right way to remember Rachel Carson". *The New Yorker*, March 26 2018.
8   Carrington, Damian. 2019. "Ocean acidification can cause mass extinction, fossils reveal". *The Guardian*, October 21, 2019. Harvey, Fiona. 2019. "Oceans losing oxygen at unprecedented rate, experts warn". *The Guardian*, December 7, 2019. Bates, N.R., Y.M. Astor, M.J. Church et al. 2014. "A time-series view of changing surface ocean chemistry due to ocean uptake of CO_2 and ocean acidification". *Oceanography* 27(1).
9   Fecht, Sarah. 2019. "Changes in the ocean 'conveyor belt' foretold abrupt climate changes by four centuries". *Earth Institute*. Accessed March 20, 2019. https://blogs.ei.columbia.edu/2019/03/20/amoc-ocean-conveyor-belt-climate-change/
10   Hylton, Wil S., 2020. "History's largest mining operation is about to begin: It's underwater – and the consequences are unimaginable". *The Atlantic*, January/February 2020.

17   Social inequality
1   Chakrabarty, Dipesh. 2009. "The climate of history: Four theses". *Critical Inquiry* 35: 197–222.
2   Malm, Andreas and Alf Hornborg. 2014. "The geology of mankind? A critique of the Anthropocene narrative". *The Anthropocene Review* 1(1): 62–69.doi: https://doi.org/10.1177/2053019613516291.
3   Chakrabarty, Dipesh. 2012. "Postcolonial studies and the challenge of climate change". *New Literary History* 43(1): 1–18. Chakrabarty, Dipesh. 2017. "The politics of climate change is more than the politics of capitalism". *Theory, Culture & Society* 34(2–3): 25–37.
4   Howe, Cymen and Anand Pandin eds. 2020. *Anthropocene Unseen: A Lexicon*. Punctum Books. Goleta: Punctum Books. Crate, Susan A. and Mark Nuttall. 2009. *Anthropology and Climate Change: From Encounters to Actions*. Walnut Creek: Left Coast Press. Wainwright, Joel and Feoff Mann. 2018. *The Climate Leviathan: A Political Theory of Our Planetary Future*. London: Verso.
5   Neate, Rubert. 2020. "Luxury travel: 50 wealthy tourists, eight countries … one giant carbon footprint". *The Guardian*, January 20, 2020.
6   Purdy, Jedediah. 2019. *This Land is Our Land: The Struggle for a New Commonwealth*. Princeton: Princeton University Press.
7   UN News. 2019. "World faces 'climate apartheid' risk, 120 more million in poverty: UN expert". United Nations. Accessed May 12 2020. https://news.un.org/en/story/2019/06/1041261
8   Grusin, Richard ed. 2017. *Anthropocene Feminism*. Minneapolis: University of Minnesota Press.
9   Singh, Ilina. 2012. "Human development, nature and nurture: Working beyond the divide". *BioSocieties* 7: 308–321.
10   Lock, Margaret and Gísli Pálsson. 2016. *Can Science Resolve the*

Nature/Nurture Debate? Cambridge:Polity Press.

18 Anthropocenes, North and South

1 Chakrabarty, Dipesh. 2012. "Postcolonial studies and the challenge of climate change". *New Literary History* 43(1): 1–18.
Thomas, Julia Adeney. 2014. "History and biology in the Anthropocene: Problems of scale, problems of value". *American Historical Review* December: 1587-1607.

2 Moore, Jason. 2013. *Anthropocene or Capitalocene? Nature, History, and the Crisis of Capitalism.* Oakland: PM Press
Tsing, Anna et al eds. 2017. *Arts of Living on a Damaged Planet: Ghosts and Monsters of the Anthropocene.* Minneapolis, MN: University of Minnesota Press.

3 Hann, Chris. 2017. "The Anthropocene and anthropology: Micro and macro perspectives". *European Journal of Social Theory* 20(1):183–196.

4 Pletsch, Carl E. 1981. "The three worlds, or the division of social scientific labor, circa 1950-1975". *Comparative Studies in Society and History* 23(4): 565–90.

5 Hecht, Gabrielle. 2018. "The African Anthropocene". *Aeon Essays.* Accessed February 6 2020. https://aeon.co/essays/if-we-talk-about-hurting-our-planet-who-exactly-is-the-we.

6 Hecht, Gabrielle. 2014. *Being Nuclear: Africans and the Global Uranium Trade.* Cambridge: The MIT Press.

7 Hecht, Gabrielle. 2018. "The African Anthropocene". *Aeon Essays.* Accessed February 6 2020. https://aeon.co/essays/if-we-talk-about-hurting-our-planet-who-exactly-is-the-we.

8 Kalmoy, Abdirashid Diriye. 2019. "The African Anthropocene: The making of a Western environmental crisis". *The New Turkey*, August 5 2019.

9 Dutt, Kuheli. 2020. "Race and racism in the geosciences". *Nature Geoscience* 13, 2–3 (2020). https://doi.org/10.1038/s41561-019-0519-z.
Goldberg, Emma. 2019. "Earth science has a Whiteness problem". *New York Times* December 23 2019.

10 Pronczuk, Monika. 2020. "How the Venice of Africa is losing its battle against the rising ocean". *The Guardian*, January 28 2020.

19 The sixth mass extinction

1 Purvis, Andy. n.d. "A million threatened species? Thirteen questions and answers". IPBES. Accessed August 19 2019. https://www.ipbes.net/news/million-threatened-species-thirteen-questions-answers.

2 Kolbert, Elizabeth. 2014. *The Sixth Extinction: An Unnatural History.* New York: Henry Holt & Company.

3 Simmons. I.G. 1989. *Changing the Face of the Earth: Culture, Environment, History.* Oxford: Basil Blackwell.

4 Moynihan, Thomas. 2019. *Spinal Catastrophism: A Secret History.* Cambridge: MIT Press.

5 May, Todd. 2018. "Would Human Extinction Be a Tragedy?" *New York Times* December 17 2018.

6 van Dooren, Thom. 2014. *Flight Ways: Life and Loss at the Edge of Extinction.* New York: Columbia University Press.

7 Rose, Deborah Bird. 2013. "Slowly – writing into the Anthropocene". In *TEXT Special Issue 20: Writing Creates Ecology and Ecology Creates Writing*, edited by Martin Harrison et al. http://www.textjournal.com.au/speciss/issue20/content.html.

8 Center for Biological Diversity, Arizona. Endangered Species Condoms. Accessed 20 August 2019. https://www.endangeredspeciescondoms.com/condoms.html.

20 Ignorance and denial

1 McGoey, Linsey. 2019. *The Unknowers: How Strategic Ignorance Rules the World.* London: Zed Books.

2 Carson, Rachel. 2018. *Silent Spring and Other Environmental Writings.* New York: Library of America.

3 Proctor, Robert N. 2012. *Golden Holocaust: Origins of the Cigarette Catastrophe and the Case for Abolition.* Berkeley: University of California Press.

4 Maslin, Mark. 2019. "Here are five of the main reasons people continue to deny climate change". *The Conversation*, November 30 2019.

5 Feik, Nick. 2020. "A national disaster". *The Monthly*, January 7 2020.

6 Samios, Zoe and Andrew Hornery. 2020. "'Dangerous, misinformation': News Corp employee's fire coverage email". *The Sydney Morning Herald.* January 10 2020.

7 Griffiths, Tom. 2020. "Savage summer". *Inside Story*, January 8 2020.

8 Williamson, Bhiamie, Jessica Weir, and Vanessa Cavanagh. 2020. "Strength from perpetual grief: How Aboriginal people experience the bushfire crisis". *The Conversation*, January 10 2020.

9 Wikipedia. n.d. "Björn Lomborg". Accessed January 10, 2020. https://en.wikipedia.org/wiki/Björn_Lomborg.

10 Lomborg, Bjørn. "Thought control". *The Economist*, January 9 2003.

11 Krugman, Paul. 2020. "Australia shows us the road to hell". *New York Times*, January 9 2020.

12 Knaus, Christopher. 2020. "Disinformation and lies are spreading faster than Australia's bushfires". *The Guardian*, January 11 2020.

13 Maslin, Mark. 2019. "Here are five of the main reasons people continue to deny climate change". *The Conversation*, November 30 2019.

14 Milman, Oliver. 2020. "Revealed: Quarter of all tweets about climate crisis produced by bots". *The Guardian*, February 21 2020.

15 Wagner, Wendy, Elizabeth Fisher, and Pasky Pascual. 2018. "Whose science? A new era in regulatory science wars". *Science* 362(6415): 636-639.
Plumer, Brad and Coral Davenport. 2019. "Science under attack: How Trump is sidelining researchers and their work". *New York Times* , December 28 2019.

21 Down to Earth

1 Gurevich, Aron. 1992. *Historical Anthropology of the Middle Ages.* Edited by J. Howlett. Oxford: Polity Press.

2 Panowski, Erwin. 1991. *Perspective as Symbolic Form.* Translated by Christopher S. Wood. London: Zone Books.

3 Gudeman, Stephen and Alberto Rivera. 1990. *Conversations in Columbia: The Domestic Economy in Life and Text.* Cambridge: Cambridge University Press.

4 Halperin, Ilana. 2013. "Autobiographical trace fossils". In *Making the Geologic Now: Responses to Material Conditions of Contemporary Life*, edited by Elizabeth Ellsworth and Jamie Kruse. Brooklyn: Punctum Books.

5 Bennett, Jane. 2010. *Vibrant Matter: A Political Ecology of Things.* Durham: Duke University Press.

6 Margulis, Lynn and Dorion Sagan. 1995. *What is Life?* Berkeley: University of California Press.

7 Pálsson, Gísli and Heather Anne Swanson. 2016. "Down to Earth: Geosocialities and Geopolitics". *Environmental Humanities* 8(2): 149–171.

8 Waldman, Johnny. 2016. "The Japanese museum of rocks that look like faces". *Colossal.* Accessed January, 2020. https://www.thisiscolossal.com/2016/11/the-japanese-museum-of-rocks-that-look-like-faces/

22 The case of freezing lava

1 Marsh, George Perkins. 1965. *Man and Nature.* Edited David Lowenthal. Cambridge: Harvard University Press.

2 Pálsson, Gísli and Heather Anne Swanson. 2016. "Down to Earth: Geosocialities and Geopolitics". *Environmental Humanities* 8(2): 149–171.
Pálsson, Gísli. 2020. *Down to Earth: A Memoir.* Goleta: Punctum Books.

3 McPhee, John. 1989. "Cooling the lava". In *The Control of Nature*, 95-179. New York: Farrar Straus Giroux.

4 McPhee, John. 1989. "Cooling the lava". In *The Control of Nature*, 95-179. New York: Farrar Straus Giroux.

23 Missed opportunities

1 Cronon, William. 1996. *Uncommon Ground: Rethinking the Human Place in Nature.* New York: W.W. Norton & Company.

2 Langematz, Ulrike. 2019. "Stratospheric ozone: Down and up through the anthropocene". *ChemTexts* 5: 1–8.

3 Blakemore, Erin. 2016. "The ozone hole was super scary, so what happened to it?" *Smithsonian Magazine*, January 13 2016
Grinspoon, David. 2016. *Earth in Human Hands: Shaping Our Planet's Future.* New York: Grand Central Publishing.
Harper, Charles L. 1996. *Environment and Society: Human Perspectives on Environmental Issues.* Upper Saddle River: Prentice Hall.

4 Skydstrup, Martin. 2016. "Trickled or troubled natures? How to make sense of 'climategate.'" *Environmental Science & Policy* 28: 92–99.
Sheppard, Kate. 2011. "Climategate: What really happened?" *Mother Jones*, April 21 2011.

24 Engineering Earth

1 Feshbach, Herman, Tetsuo Matsui, and Alexandra Oleson eds. 2014. *Niels Bohr: Physics and the World.* London: Routledge.

2 Masco, Joseph. 2010. "Bad weather: On planetary crisis". *Social Studies of Science.* 40(1): 7-70.

3 Hawken, Paul ed. 2017. *Drawdown: The Most Comprehensive Plan Ever Proposed to Reverse Global Warming.* London: Penguin.

4 Harper, Charles L. 1996. *Environment and Society: Human*

Perspectices on Environmental Issues. Upper Saddle River: Prentice Hall.

5   Pettifor, Ann. 2019. *The Case for the Green New Deal.* London: Verso Klein, Naomi. 2019. *On Fire: The (Burning) Case for a Green New Deal.* New York: Simon and Schuster.
      Purdy, Jedediah. 2019. *This Land is Our Land: The Struggle for a New Commonwealth.* Princeton: Princeton University Press.

6   Buck, Holly Jean. 2019. *After Engineering: Climate Tragedy, Repair, and Restoration.* London: Verso.

7   Macfarlane, Robert. 2018. *Underland: A Deep Time Journey.* London: Penguin
      Magnason, Andri Snær. 2020. *On Time and Water.* In press.

8   Frisch, Max. 1980. *Man in the Holocene: A Story.* London: Dalkey Archive Press.

9   Zarin, Cynthia. 2015. "The artist who is bringing icebergs to Paris". *The New Yorker,* December 2015.

10   Ckakrabarty, Dipesh. 2009. "The climate of history: Four theses". *Critical Inquiry* 35(2): 197–222
      Santana, Carlos. 2019. "Waiting for the Anthropocene". *Brit. J. Phil. Sci* 70: 1073–1096.
      Clark, Nigel and Bronislaw Szerszynski. 2020. *Planetary Social Thought: The Anthropocene Challenge to the Social Sciences.* Oxford: Polity Press.

11   Arendt, Hannah. 1958. *The Human Condition.* Chicago: University of Chicago Press.

12   Szerszynski, Bronislaw. 2003. "Technology, performance, and life itself: Hannah Arendt and the fate of nature". *Sociological Review* 51: 2013–218
      Pálsson, Gísli, Bronislaw Szerszynski, Sverker Sörlin et al. 2013. "Reconceptualizing the 'Anthropos' in the Anthropocene: Integrating the social sciences and humanities in global environmental change research". *Environmental Science and Policy* 28: 3–13.

25   Fixing carbon and buying time

1   Elder, W. A. 2014. "Drilling into magma and the implications of the Icelandic Deep Drill Drilling Project (IDDP) for high-temperature geothermal systems worldwide". *Geothermics* 49: 111-118.

2   Clark, Nigel, Alexandra Gormally, and Hugh Tuffen. 2018. "Speculative volcanology: Time, becoming, and violence in encounters with magma". *Environmental Humanities* 10(1): 274-294.

3   CarbFix. n.d. Accessed January 24, 2020. https://www.carbfix.com.

4   Snæbjörnsdóttir, Sandra Ó, Bergur Sigfússon, Chiara Marieni et al. 2020. "Carbon dioxide storage through mineral carbonation". *Nature Reviews,* January 20 2020. doi: doi.org/10.1038/s43017-019-0011-8.

5   Gislason, Sigurdur and Eric H. Oelkers. 2014. "Carbon storage in basalt". *Science* 344 (April 25): 373–374.

6   Gunnarsson, Ingvi, Edda S. Aradóttir, Eric H. Oelkers et al. 2019. "The rapid and cost-effective capture and surface mineral storage of carbon and sulfur at the CarbFix2 site". *International Journal of Greenhouse Gas Control* 79: 117-126.

7   Snæbjörnsdóttir, Sandra Ó, Bergur Sigfússon, Chiara Marieni et al. 2020. "Carbon dioxide storage through mineral carbonation". *Nature Reviews,* January 20 2020. doi: doi.org/10.1038/s43017-019-0011-8.

8   Big Sky Carbon Sequestration Partnership. "Basalt Pilot Project". Accessed January 24, 2020. https://www.bigskyco2.org/research/geologic/basaltproject.

9   Snæbjörnsdóttir, Sandra Ó, Bergur Sigfússon, Chiara Marieni et al. 2020. "Carbon dioxide storage through mineral carbonation". *Nature Reviews,* January 20 2020. doi: doi.org/10.1038/s43017-019-0011-8.

10   Snæbjörnsdóttir, Sandra Ó, Bergur Sigfússon, Chiara Marieni et al. 2020. "Carbon dioxide storage through mineral carbonation". *Nature Reviews,* January 20 2020. doi: doi.org/10.1038/s43017-019-0011-8.

11   This condensed account is partly based on an interview (January 23 2020) with one of the architects of the CarbFix project, Sigurður Reynir Gíslason, professor of geochemistry at the University of Iceland.

26   Rebellion

1   Walls, Laura Dassow. 2017. *Henry David Thoreau: A Life.* Chicago: The University of Chicago Press.

2   Walls, Laura Dassow. 2017. *Henry David Thoreau: A Life.* Chicago: The University of Chicago Press.

3   Extinction Rebellion. 2019. *This is Not a Drill: An Extinction Rebellion Handbook.* London: Penguin.

4   Hunziker, Robert. 2019. "Extinction Rebellion: What is it?" *countercurrents.org.* Accessed September 7 2019. https://countercurrents.org/2019/09/extinction-rebellion-what-is-it
      Mackintosh, Eliza. 2019. "A psychedelic journey, a radical strategy and perfect timing". CNN, December 25 2019.

5   Interview with Cambridge activist, anthropologist Sarah Abel. October 2019.

6   Berg, Nate. 2020. "'We want a new Mayor!': Inside the Berlin city game for children". *The Guardian,* January 1 2020.

7   Alter, Charlotte, Suyin Haynes, and Justin Worland. 2019. "Time person of the year 2019: Greta Thunberg". *Time,* December.

27   Geopolitics as housekeeping

1   Wendt, Wolf Rainer. 2018. "House, state, and world fundamental concepts of societal governance in the West and East in comparison". *Asian Journal of German and European Studies* 3 (11): 1–13.

2   Hollingshead, A. B. 1940. "Human ecology and human society". *Ecological Monographs* 10(3): 354–366.

3   Pálsson, Gísli and Heather Anne Swanson. 2016. "Down to Earth: Geosocialities and geopolitics". *Environmental Humanities* 8(2): 149–171.

4   Dalby, Simon. 2014. "Rethinking geopolitics: Climate security in the Anthropocene". *Geopolicy* 5(1): 1-9.

5   Ernman, Malena, Greta Thunberg, Beata Thunberg, and Svante Thunberg. 2018. *Our House is on Fire: Scenes of a Family and a Planet in Crisis.* London: Penguin Books.

6   Prainsack, Barbara. 2017. *Personalized Medicine: Empowered Patients in the 21st Century?* New York: New York University Press
      Prainsack, Barbara and Alena Buyx. 2011. *Solidarity: Reflections on an Emerging Concept in Bioethics.* London: Nuffield Council of Bioethics.

7   The New Stone Age. 2020. Accessed 6 March 2020. https://www.buildingcentre.co.uk/whats_on/exhibitions/the-new-stone-age-2020-02-27#part-of-series.
      Wainwright, Oliver. 2020. "The miracle new sustainable product that's revolution architecture – Stone!" *The Guardian,* March 4 2020.

8   Gibson-Graham, J.K. 2011. "A feminist project of belonging for the Anthropocene". *Gender, Place & Culture* 18(1): 1–21
      Prattico, Emilie. 2019. "Habermas and climate action". *Aeon,* December 18 2019.

9   Pettifor, Ann. 2019. *The Case for the Green New Deal.* London: Verso
      Klein, Naomi. 2019. *On Fire: The (Burning) Case for a Green New Deal.* New York: Simon and Schuster
      Purdy, Jedediah. 2019. *This Land is Our Land: The Struggle for a New Commonwealth.* Princeton: Princeton University Press.

10   Lenton, Timothy M. Johan Rockström, Owen Gaffney et al. 2019. "Climate tipping points – too risky to bet against". *Nature* 575: 592–595.

11   Leahy, Stephen. 2019. "Climate change driving entire planet to dangerous 'tipping points'". *National Geographic.* November 27 2019.

12   Connolly, William E. 2017. *Facing the Planetary: Entangled Humanism and the Politics of Swarming.* London: Duke University Press.

28   A fuming planet

1   Dentith, Simon. 1995. *Bakhtinian Though: An Introductory Reader.* London: Routledge.

2   Lenton, Timothy M. Johan Rockström, Owen Gaffney et al. 2019. "Climate tipping points – too risky to bet against". *Nature* 575: 592–595.

3   Linden, Eugene. 2019. "How scientists got climate change so wrong". *The New York Times,* November 8 2019.

4   Degroot, Dagomar. 2018. *The Frigid Golden Age: Climate Change, the Little Ice Age, and the Dutch Republic, 1560–1720.* Cambridge: Cambridge University Press.

5   Degroot, Dagomar. 2019. "Little Ice Age lessons". *Aeon,* November 11 2019.

INDEX

· · · · · · · · · · · · · · · · · · · ·

ACKNOWLEDGEMENTS

I thank the Icelandic Centre for Research (Rannís) and the University of Iceland Research Fund for funding my research on the Anthropocene and related issues for the last two decades or so. During my research I benefited from short stays at the Centre for Advanced Study in Oslo (CAS) and the anthropology departments of the University of Cambridge and the University of Copenhagen. Numerous people – friends, colleagues, collaborators, translators, mentors and other significant others – deserve credit for one thing or another, drawing my attention to important sources, stimulating exciting discussions, and developing ideas which helped me, directly or indirectly, to assemble and craft this text; most notably (a full list would be much too long) Sarah Abel, Dominic Boyer, Nancy Marie Brown, Finnur Ulf Dellsén, Kathrina Downs-Rose, Paul Durrenberger, Níels Einarsson, Sigurður Reynir Gíslason, Stephen Gudeman, Guðný S. Guðbjörnsdóttir, Sigurður Örn Guðbjörnsson, Ari Trausti Guðmundsson, Kirsten Hastrup, Ilana Halperin, Shé Hawke, Karen Holmberg, Edward H. Huijbens, Jón Haukur Ingimundarson, Tim Ingold, Valdimar Leifsson, Marianne Elisabeth Lien, Örn D. Jónsson, Bonny McCay, Sarah Keene Meltzoff, Astrid Ogilvie, Andri Snær Magnason, Barbara Prainsack, Elspeth Probyn, Hugh Raffles, Heather Anne Swanson, Cymene Howe, Bronislaw Szerszynski, Sverker Sörlin, Thom van Dooren, Hendrik Wagenaar and Anna Yates. Finally, I thank Welbeck and its staff, in particular Isabel Wilkinson, for the opportunity to engage with this book; their encouragement and guidance in structuring the work and handling the production process have been extremely valuable.

Opposite: *Scenic view of sunrise in Grand Canyon national park, Arizona.*

223
ACKNOWLEDGEMENTS

CREDITS

．．．．．．．．．．．．．．．．．．．．．

The publishers would like to thank the following sources for their kind permission to reproduce the pictures in this book.

Alamy: Olivier Asselin 135; /Classicpaintings 209; /Danita Delimont Creative 51 l; /Bert de Ruiter 31; /GL Archive 210-211; /Granger Historical Picture Archive 22 r; /History and Art Collection 40 r; /Billy H.C. Kwok 56 t; /National Geographic Image Collection 118; /Newscom 157; /Science History Images 61; /Universal Images Group North America LLC 15; /World History Archive 53 b

Hugo Ahlenius, UNEP: 146

Helle Astrid: 28

Author photograph: 93, 110

Bridgeman Images: 23 l, 39; /Look and Learn 50; /The Stapleton Collection 40 l

British Library: 21

CarbFix: 188; /Photo by Arni Saeberg 190-191; /Photo by Sandra O Snaebjornsdottir 189

Chong Chen, Katsuyuki Uematsu, Katrin Linse & Julia D. Sigwart: 121 tl

Collectie Menno Huizinga: 127

© The Economist/Copyright Clearance Center: 8

www.endangeredspeciescondoms.com: 145

Getty Images: AFP 121ltr; /Tolga Akmen/AFP 142-143, 193t, 194-195; /Michael Appleton/NY Daily News Archive 68; / Bettmann 161; /Michael Campanella 197; /Jana Cavojska/ SOPA Images/LightRocket 136, 137; /Chesnot 183; /Giles Clarke 126; /Fabrice Coffrini 94-95; / Ashley Cooper/Corbis 119t; /DeAgostini 36; /Pius Utomi Ekpei/AFP 130-131; /Sean Gallup 12; /David Gray 150-151; /Signy Asta Gudmundsdottir 107; /Halldor Kolbeins/AFP 184-185; /Brett Hemmings 153; /Fred Ihrt/LightRocket 160, 165, 166; /Chris Jackson 96; /Dan Kitwood 92; /Michael Kraus/EyeEm 132-133; /Joao Laet/AFP 33; /Library of Congress/Corbis/VCG 43; /Win McNamee 99; /Nick Moir/The Sydney Morning Herald 33-34; /Filippo Monteforte/AFP 89; /Loen Neal 193 b; /Justin Paget 66; /Wolter Peeters/The Sydney Morning Herald 148; /Ashley Pon 76; /Print Collector 54; /Joe Raedle 100-101; /Arni Saeberg/Bloomberg 112-113; /Mujahid Safodien/AFP 134; /Oli Scarff 129; /Sarah Silbiger 168; /SIPA 186-187; / Guy Smallman 196; /Yuri Smityuk/TASS 9, 140-141; /Chip

Somodevilla 175; / SSPL 108-109; /Stringer/AFP 124-125; / Sygma 62-63; /Charly Triballeau/AFP 85; /William West/ AFP 206-207; /Ed Wray 178-179

Courtesy of Ilana Halperin: 156

Lagos Tomorrow' - Water Cities concept by NLÉ. Image by NLÉ in collaboration with Lekan Jeyifous: 57

Nick Longrich: 31

Mary Evans Picture Library: 46 l

WilliamJ. Mitschetal: 69

National Archives & Records Administration: 147

NASA: 65, 83, 119b, 172, 201, 204

National Archives of Iceland: 71

Nature Picture Library: Jordi Chas 74; /Michael Pitts 139; / Tony Wu 75

New York Public Library: 37 b

Sveinn Þormóðsson.Brynja Pétursdóttir's Private Collection: 167

© Pekka Niittyvirta and Timo Aho: 91

NOAA: 102

The Ohio State University, ORWRP: 67

Private Collection: 19, 20, 23, 24-25, 42 t, 48, 59, 79, 111, 116, 154, 155 b

Gregg Segal Photography: 78

Shutterstock: 46 r, 58; /Best Works 120; /Bob63 202; / Nic Bothma/EPA-EFE 73; /Ronnie Chua 98; /Martin 303 223; /Marcel Derweduwen: 86-87; /Iacomino Frimages 14; /Peter Hermes Furian 90; /HBO/Kobal 64; /Holli 198-199; /Rene Holtslag 123; /Jer123 174; /Kedofoto 158-159; / Nicola Marfisi/AP 11; /Mopic 18; /Morphart Creation 114-115; /Anna Moskvina 47; /NOAA 102; /Eraldo Peres/AP 37t; /Satakorn 60; /Siberian Art 27; /Trekandshoot 176; / Universal History Archive/UIG 106; /Friedemann Vogel/ EPA-EFE 16; /Seth Wenig/AP 77; /Xinhua News Agency 103; /20th Century Fox/Kobal 104-105

The Science Institute, University of Iceland: 162-163